FINGERPONDS: MANAGING NUTRIENTS AND PRIMARY PRODUCTIVITY FOR ENHANCED FISH PRODUCTION IN LAKE VICTORIA'S WETLANDS, UGANDA

Promoter: Prof. dr. P. Denny
Professor of Aquatic Ecology
UNESCO-IHE/Wageningen University, The Netherlands

Co-promoters: Dr. A.A. van Dam
Senior Lecturer in Ecological and Environmental
Modelling
UNESCO-IHE, The Netherlands

Prof. dr. F. Kansiime
Director, Makerere University Institute of Environment
and Natural Resources
Kampala, Uganda

Awarding Committee: Prof. dr. M.A. Wahab
Bangladesh Agricultural University, Bangladesh

Prof. Dr. B. Moss
University of Liverpool, United Kingdom

Prof. dr. M. Scheffer
Wageningen University, The Netherlands

Dr. J.J.A. van Bruggen
UNESCO-IHE, The Netherlands

Fingerponds: managing nutrients and primary productivity for enhanced fish production in Lake Victoria's wetlands, Uganda

DISSERTATION

Submitted in fulfilment of the requirements of
the Academic Board of Wageningen University and
the Academic Board of the UNESCO-IHE Institute for Water Education
for the Degree of DOCTOR
to be defended in public
on Wednesday, 20 December 2006 at 10:30 hours
in Delft, The Netherlands

by

ROSE CHRISTINE KAGGWA
born in Tororo, Uganda

Taylor & Francis is an imprint of the Taylor & Francis Group, an informa business

© 2006, Rose Christine Kaggwa

Published by:
Taylor & Francis/Balkema
PO Box 447, 2300 AK Leiden, The Netherlands
e-mail: Pub.NL@tandf.co.uk
www.balkema.nl, www.taylorandfrancis.co.uk, www.crcpress.com

ISBN10 0-415-41697-3 (Taylor & Francis Group)
ISBN13 978-0-415-41697-9 (Taylor & Francis Group)
ISBN 90-8504-498-7 (Wageningen University)

'To my loving family,

To God be the Glory great things he hath done'.

Abstract

Wetlands are of great ecological importance and have been described as the most important zone for freshwater fisheries. They can be exploited for agriculture and aquaculture through the integration of these activities. In Sub-Saharan Africa, emphasis has been put on agriculture and this has led to the degradation of these ecosystems. With the ever-increasing population rise, many of the wetlands are encroached on and degraded. These wetlands are best managed under the framework of a 'working wetland' which ensures a rational compromise between ecological condition and the level of human utilization.

Food security in Sub-Saharan Africa is an ever-increasing problem with over 200 million undernourished people. The per capita fish consumption rate has dropped from 9 kg in 1973 to 6.7 kg in 2005. More and more people have no access to protein. In addressing the first Millennium Development Goal: reduction of poverty and hunger, low-cost strategies are sought that can ameliorate food scarcity in many parts of the region principally in the drier seasons. Fish culture in seasonal wetland fish ponds 'Fingerponds' on the shores of Lake Victoria provides a wise use option for the management of natural wetlands while at the same time improving food security for resource-poor communities.

The core objective of this dissertation is to evaluate the functioning and management of Fingerponds based on the application of natural organic manure and the use of artificial substrates. This thesis determines how nutrient dynamics and primary productivity regulate fish production in these systems and examines the underlying factors that determine the variability in pond water quality and fish production. The thesis provides suitable management practices that are easily adoptable by resource-poor communities.

Eight Fingerponds (each 192 m^2) were constructed in two villages (Gaba and Walukuba) on the northern shores of Lake Victoria, Uganda. In 2002, the wetland ecotone (i.e. ponds, wetland and lake inshore zones) characteristics were determined and their potential for fish culture ascertained. Natural stocking of ponds occurred in April/May 2003. This research demonstrates that Fingerponds provide a conducive environment for the culture of fish which is dependant on soil conditions, seasonal flooding, natural stocking, and water quality. Seasonal floods enable fish from lakes or rivers to migrate to the ponds and their levels and duration determine the fish species stocked in the ponds. An assessment of the ecological processes revealed buffering capacity and sedimentation/re-suspension as the main ecological processes determining water quality in these freshwater wetland ecosystems.

To sustain good fish growth throughout the dry season in Fingerponds, natural food production for fish was enhanced through the application of organic manure (chicken and fermented green manure) and use of artificial substrates for periphyton development. Firstly, the effects of chicken manure application on primary productivity and water quality were examined. The experiments revealed that rates of chicken manure 520 to 1563 kg ha^{-1} applied fortnightly increased phytoplankton primary productivity without adversely affecting pond water quality. Furthermore, concentrations of oxygen and ammonia were maintained within limits acceptable for fish survival and growth. The main limitations to primary productivity (phytoplankton and periphyton) in these flood-fed ponds were inorganic clay turbidity, decreasing water levels and nitrogen limitation. Secondly, use of bamboo and local wetland plants (*Raphia,* papyrus and *Phragmites*) revealed that bamboo and *Phragmites* substrates were the most suitable for periphyton development. In the presence of artificial substrates, primary productivity doubled.

This study revealed that Fingerponds can play a major role in ameliorating food security for resource-poor communities through the provision of protein (fish). A maximum fish yield of up to 2.67 tonnes $ha^{-1} yr^{-1}$ (i.e. per capita fish consumption of 6.2 kg over a 310 day growth period) was attained from periphyton-based organically manured Fingerponds. However, the study also reveals that fish growth in flood-fed Fingerponds is limited by high reproduction/recruitment of fingerlings (leading to high feeding pressure and subsequent stunting of fish) hence small sized fish; low water levels, high light limitation due to inorganic clay turbidity (hence lowered primary productivity) and low zooplankton biomass. Manual sexing of fish as a management strategy results in an increased ratio of male to female fish. Furthermore, periodic removal of female fish and fingerlings is unsuccessful in curbing reproduction.

Fingerponds can be distinctively separated into two types: typical lake-floodplain Fingerponds and those fed also with inflowing rivers in this case referred to as river floodplain Fingerponds. A dynamic model was used to simulate fish growth in these ponds. The model was able to capture the dynamics of hydrology, nutrients and fish in the two types of Fingerponds, demonstrating that similar fundamental processes underlie fish production in these systems. Model fish yields of up to 2800 kg ha^{-1} and water quality predictions were comparable to field measurements. Using the model, nitrogen budgets for Fingerponds were calculated and quantitative estimates of all process flows were given. The model though preliminary once subjected to sensitivity analysis and validation can be used as a management tool. Main knowledge gaps pertain to light limitation of primary productivity and the food selectivity of tilapia and their fingerlings. In its current state, the model is a research tool that identifies knowledge gaps and can be applied to frame hypotheses for further applied research. Ultimately, the model can contribute to improving the management of Fingerponds.

Finally in conclusion, the research has demonstrated that with correct management, enhanced fish production in Fingerponds is possible.

Table of Contents

Chapter 1

The role of natural wetlands in enhancing food security - a general introduction

Introduction

This dissertation evaluates the functioning and management of seasonal wetland fish ponds 'Fingerponds' in Uganda, based on the application of natural organic manure and the use of artificial substrates. It determines how nutrient dynamics and primary productivity regulate fish production in these systems. The underlying factors that determine the variability in pond water quality and fish production are analyzed. This chapter considers food security in Sub-Saharan Africa; outlines the role and potential for fish culture in wetlands as a means to ameliorate food scarcity and highlights the status of aquaculture in the Sub-Saharan African region. The Fingerpond concept is defined and the criteria for Fingerpond site selection, design and construction are given. The gaps in knowledge are highlighted. A summary of the characterization of the study wetlands are highlighted. Lastly, the research strategy, scope, objectives and outline of the thesis are given.

Food security and nutrition in Sub Saharan Africa – the challenges

Good nutrition is the cornerstone for survival, health and development for current and future generations (UNICEF, 2001). The first Millennium Development Goal (MDG) addresses eradication of extreme poverty and hunger. One of its two targets is to reduce by half the population of people who suffer from hunger by 2015 (UN, 2005). Over 800 million people in the world today have too little to eat to meet their daily energy needs. A third of these are in Sub-Saharan Africa (SSA) and are affected by malnutrition (Benson, 2004; UN, 2005). Subsequently, in Sub-Saharan Africa, this accounts for more than half of children deaths with most of the children not able to obtain adequate protein or vitamins (UNICEF, 2005). Food security and nutrition play a major role in achievement of the fourth and sixth MDG goals; reduction of child mortality and combating of HIV/AIDS, malaria and other diseases by 2015. Health, food security and nutrition are intertwined and cannot be addressed in isolation of each other.

Food security is a universal goal. However, the limits to a sustainable level of food production are set by the availability of resources (e.g. land, water) and human capacity to increase productivity of these resources without depleting or degrading them. Although fish accounts for 20% of all the animal protein in the human diet (Williams, 1996), in Africa, the average fish consumption continues to decline (Figure 1.1). There is therefore a need to increase food security especially in the dry season when livelihoods are most at risk (Denny et al., 2006).

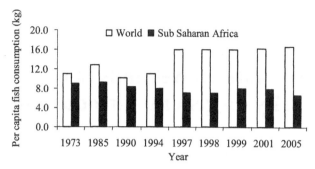

Figure 1.1 Trends in per capita fish consumption in the World and Sub-Saharan Africa (Source: Delgado *et al.*, 2003; FAO, 2000, 2004a; WFC, 2005).

With Africa's population growth rate at an average 2.5 % (WDI, 2005), the pressure on natural resources has increased. This is due to changes in hydrological regimes (such as dam construction), over-exploitation and environmental degradation from pollution, eutrophication and the introduction of exotic species, amongst other reasons. Consequently, once very productive lakes such as Lake Victoria are now showing a steady decline in fisheries (Witte *et al.*, 1995; Geheb & Binns, 1997). The ever-increasing international demand for fish is taking its toll on both large and small-scale fisheries. The high demand for fish has led to some profound and disastrous consequences such as the illicit poisoning of fish to increase catches. Competition for fish in the absence of appropriate management continues to result in rapid depletion of resources, the destruction of ecosystems and fish habitats and diminishing economic returns. In the midst of competition, ignorance of existing regulations and control mechanisms, cases of illegal, unreported and unregulated fishing practices arise (Drammeh, 2001). Emphasis is put on economic benefits but these often do not trickle down to the poor and needy communities. The consequent ecological decline leads to poverty, and poverty to ecological decline in a vicious cycle that can be tackled only through the development and implementation of enlightened policies.

Wetland ecotones and their role in fisheries

Wetlands are defined by the Ramsar Convention as "areas of marsh, fen, peat land or water, whether natural or artificial, permanent or temporary, with water that is static or flowing, fresh, brackish or salt, including areas of marine water the depth of which at low tide does not exceed six metres". In addition, "wetlands may incorporate adjacent riparian and coastal zones, islands or bodies of marine water deeper than six metres at low tide lying within the wetlands". This includes habitat types like rivers, lakes, coastal lagoons, mangroves, peat lands and coral reefs, as well as human-made wetlands such as fish and shrimp ponds, farm ponds, irrigated agricultural land (including rice fields), salt pans, reservoirs, gravel pits, sewage farms and canals'(RCS, 2006)

Wetland ecotones are zones of transition between adjacent ecological systems, having sets of characteristics uniquely defined by space and time scales and by the strength of interactions between adjacent ecological systems (Holland, 1988). Water fluctuations determine the structural variation (both in time and space) of aquatic macrophytes and algae in the border zones of water bodies (Bugenyi, 1991).

The physical potential of inland valley swamps in Sub-Saharan Africa has been conservatively estimated at 135 million hectares with only 1.3% under cultivation (FAO-SAFR, 1998). In East and South Africa alone, wetland coverage varied between 3 and 10 % of the land surface. As such, wetlands provide a stable environment and social development avenue for local communities living within their vicinity. The livelihood options provided by rivers, lakes, floodplains and wetland systems include: drinking and irrigation water, crop production, livestock husbandry, fisheries, construction materials, wild life habitat and medicinal herbs.

Wetlands are of great ecological importance and are probably the most important zone for inland freshwater fisheries (Denny, 1985). They support a large invertebrate fauna, act as a feeding ground for young and growing fish and provide refugia against predators (Denny, 1985; Balirwa, 1998). They have been described as a typical habitat for tilapia species particularly in the East African lakes (Kudhongania & Cordone, 1974; Lowe-McConnell, 1975). The expansion of fisheries into the wetland areas and floodplains is an avenue for increased yields in African waters and can be exploited further through fish farming (Welcomme, 1979; Toews & Griffith, 1979; Dugan et al., 2004; Halwart & van Dam, 2006). Seasonal changes of the wetland ecotones in discharge, nutrient concentrations, pH, temperature and dissolved oxygen in turn influence the composition and abundance of the plant and animal communities inhabiting these ecosystems.

With the multiplicity of wetland functions there is a need to use wisely the interactions (social, physical, hydrological, chemical and biological) that determine the benefits of these resources. The Ramsar Convention definition of wetland wise use emphasizes the need to maintain the natural properties of the ecosystem while exploiting the benefits. However, the trade-off between environment protection and development is most acute in these fragile ecosystems.

Wetlands can be better managed under the framework of a 'working wetland', defined as "a managed wetland in which a rational compromise is made between ecological condition and the level of human utilization" (McCartney et al., 2005). This approach underpinned by the concept of 'wise use' is based on a form of multi-criteria analysis that integrates biophysical and social economic aspects of wetland utilization. This approach has been tested for agriculture but there is a need to test it in integrated agriculture-aquaculture (IAA) systems too.

Formulation and development of policies with regard to food security, fishery and wetland management requires integration of activities. Concerted efforts need to be made to ensure that policies, strategies and mechanisms aimed at poverty reduction recognize the importance of sustainable wetland management (WI, 2006). Sustainable utilization of wetlands (wise use) is utilization for the benefit of mankind in a way compatible with the maintenance of the natural properties of the ecosystem (RCS, 2004). This is best achieved through networking and partnerships with stakeholders at all fora.

Aquaculture in Sub-Saharan Africa

Earthen ponds are the dominant aquaculture production units in Sub-Saharan Africa. The location of fish ponds is governed by water availability amongst many other factors. Most of the fish ponds average between 100 and 1000 m^2. A few private larger scale commercial fish farms range between 2 to 30 ha, found mostly in Kenya, Malawi, Nigeria, Zambia and Zimbabwe. Average net annual yields from these systems range from 10 kg 100 m^{-2} (subsistence) to 30 kg 100 m^{-2} (small-scale commercial) (Coche *et al.*, 1994). Most of the produce is consumed directly by the farmer or bartered/sold within the local community. The systems require substantial investment as fingerlings and feeds have to be bought in addition to the initial construction costs. Lack of relevant research has hampered further the development of aquaculture systems. As a result, adoption of aquaculture has remained a minor priority amidst what is seen as more burning issues such as public health and education (FAO, 2004b).

Fingerponds

The concept

Fringe wetlands are inundated during wet seasons and dry out during the drier periods. As the floods recede, fish are trapped in shallow pits within the swamp. In many instances these fishes are not utilized. While in most countries of Sub-Saharan Africa wetland fringes or flood plains are used for agriculture (crop and livestock), the aquaculture potential in these swamps is under-exploited. Farmers tend to concentrate on monoculture systems, i.e. growing of either crops or fish but not both. This results in high production costs and is risky. If the crop fails or there is a fish kill, everything is lost for that season. The rationale of integrated farming is to minimize waste, encourage low costs of production, increase income, reduce risk and improve management. In management of wetland systems there is need to shift from this paradigm of 'single utilization' monoculture to integrated systems. This can contribute to better management and exploitation of wetland systems.

Fingerponds are earthen ponds dug from the landward edge of wetlands, extending like fingers into the swamp (Denny & Turyatunga, 1992) (Figure 1.2). They exploit two phenomena: (a) the high natural productivity of the swamp/lake interface (which provides the fish stock for the ponds) and (b) the trapping of fish in depressions following flood retreat on floodplains after seasonal rains. Hence the ponds are self-stocking and obviate the necessity of purchasing fingerlings which can be prohibitive for the poorest people. Soil dug from the ponds forms raised beds between the fingers and can be used for cultivation of seasonal crops. The rich organic bottom soil removed from the ponds when they dry up or are drained for harvest, acts as manure for the beds. Fingerponds can be described as agro-piscicultural systems modified from schemes developed in Mexico, China, Asia and more recently Africa (Denny *et al.*, 2006). Fish production is stimulated with natural organic manures (both from animals and plants) supplied by the local communities. This "wise use" technology retains the functions of the wetland and can go a long way to ameliorate the poor food security in many parts of Africa.

(a) Schematic drawing of Fingerponds

(b) Fingerpond site   (c) Vegetable plot on raised bed

Figure 1.2. Drawing of Fingerponds (a) schematic drawing adopted from Denny and Turyatunga, 1992 (b) Fingerpond site in Walukuba, Jinja (c) vegetable plots on raised beds in Walukuba, Jinja. Photographs by R.C.Kaggwa and F.Kansiime.

Fingerponds: a solution to poor food security and sustainable wetland management

Research on aquaculture systems has focused on enhancement of fish production through the application of organic and inorganic manure and supplementary feeds (Delincé, 1992; Egna & Boyd, 1997). More recently, enhancement of periphyton as a natural fish food through the introduction of artificial substrates has been explored in Asia and West Africa (Azim *et al.*, 2005). Periphyton-based systems provide a cheaper option for provision of natural food to the fish. However, there is exiguity in data, particularly in relation to tropical systems in Sub-Saharan Africa.

Fish culture development in the Sub-Saharan Africa region is at a cross-road. In many instances, unsustainable development has led to short-term and medium-term profits at the expense of long-term ecological balance and social stability. Less priority has been given to environmental research or to improving traditional systems such as the brush parks used in Malawi or culture-based enhancement in floodplains or wetlands (Jamu & Ayinla, 2003).

Integrated agriculture-aquaculture systems use low levels of inputs and are considered semi-intensive (FAO, 2001). They do no rely on heavy feed and fertilizer inputs and operate in synergy with agriculture (crop-livestock-fish integrated farming). They capitalize on *in situ* provision of pond inputs, e.g., animal and green manure, similar to ecological/organic aquacultural systems as described by Costa-Pierce (2002).

Fingerpond systems focus on the development of small-scale farming. They preserve the wetland environment in which they are situated while maintaining a productive culture system (Kipkemboi, 2006). They can be incorporated into sustainable fisheries management and provide an alternative model of aquaculture research and development. Amongst the main principles for ecological aquaculture are the preservation of the form and function of natural resources; trophic levels efficiency (using animal and green manures rather than supplementary feeds); and integration of the system in food production (Costa-Pierce, 2002). Locally available bio-resources (such as animal manure, farm wastes, composted kitchen wastes, cassava dough, leaves of cocoyam, cassava, papaya and cabbages) have been used successfully as pond inputs in a number of studies (e.g. Brummett & Noble, 1995; Prein *et al.*, 1996; Hishamunda *et al.*, 1998). Fingerponds provide an option for wise use of natural resources whilst providing a source of livelihood to resource-poor persons.

Gaps in knowledge

In a traditional pond with very limited substrates for periphyton development, organic manures boost fish yields by increasing phytoplankton primary productivity (Colman & Edwards, 1987). Inorganic nutrients released into the water column are taken up by the phytoplankton and contribute, through mortality, organic matter to the detritus/sludge layer at the pond bottom. Phytoplankton and bottom detritus serve as food for fish but their utilization depends on feeding preferences and morphological adaptations of the fish species stocked. When additional substrates for periphyton attachment are installed, inorganic nutrients follow the periphyton loop, adding a third food resource to the pond. Periphytic microorganisms can be consumed by fish while the dead periphyton contributes to the detritus mass in the pond as well as provides a source of food for heterotrophs (Azim *et al.*, 2005).

Fingerponds systems are different from most conventional aquaculture systems in that they are entirely dependant on natural flood events. They thus have seasonal functional periods as water levels are maintained either by precipitation or underground water infiltration during the culture period. In order to stimulate fish production in these systems, the use of natural, organic manures is proposed. The addition of organic manures to fish ponds results in enhanced algal development as shown in various studies (Delincé, 1992; Pechar, 2000). It is very important to control and monitor manure application rates. With under-fertilization, algae will not attain optimal net productivity while over-fertilization can cause oxygen depletion and self shading (Knud-Hansen, 1998). For Fingerpond systems, good management practices to regulate pond inputs and/or conditions are essential: none of this information is available. Furthermore there is need for perceived simplicity or low potential risk, particularly when targeting resource-poor communities. It is imperative that in new technologies such as Fingerponds, these factors be critically considered.

Criteria for site selection and study area location

Fingerpond research sites were selected based on the following criteria: topography, accessibility, soil type, flood history and presence of fish in the wetland. Other considerations were the land tenure and availability of a local community group. Reconnaissance visits were made to each proposed site with a team that comprised a surveyor, hydrologist, civil engineer, and fish scientist. Formal and informal meetings were held with local people within a 2 km radius of the sites. Information on the activities carried out in the wetland was obtained.

Eight Fingerponds (24x8 m, 1.5 m average depth) were constructed in two localities (Gaba and Walukuba) within the littoral region of the northern shores of Lake Victoria, Uganda (Figure 1.3). The Gaba wetland lies on the northern shores of Lake Victoria adjacent to the Inner Murchison bay (13 km South East of Kampala city centre, capital of Uganda). It is set at latitude 0° 15'N. It covers an area of 1.43 km^2. The Walukuba wetland borders the Napoleon Gulf in Jinja between latitudes 0° 22'N and 0° 30'N and longitude $22^\circ10$'E and 33° 26'E and covers an area of 3 km^2. Both wetlands lie at an altitude of 1143 m above mean sea level. This research was part of a five year 'EU INCO-DEV Fingerponds project' implemented in East Africa. Four similar sites were set up on the shores of Lake Victoria in Kusa and Nyangera (Kenya) and in the floodplains of the River Rufiji in Uba and Ruwe (Tanzania).

Design criteria of Fingerponds
A preliminary hydrological survey was carried out that involved determination of the ground water table and soil proliferation. The summarized design criteria were:

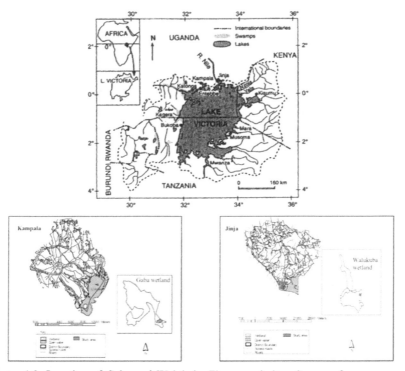

Figure 1.3. Location of Gaba and Walukuba Fingerpond sites. Source of maps: map of Uganda modified from Balirwa, 1998 used with permission, maps of Kampala and Jinja: Wetlands Inspection Division, GIS database used with permission.

a) floods to reach at least 5 m distance from the shallow end of the pond (to allow filling of ponds using a channel in case of low flood levels); b) soils to contain clay content of more than 20 % (i.e. reduced water losses through seepage) and c) accessibility (to allow easier monitoring of ponds).

Construction aspects

The Fingerpond site occupied a total area of about 500 m^2 and comprised of four ponds (each about 24 8 m, 1 m deep at the shallow end and 2 m at the deep end) and three or four raised beds (each at least 24x10 m). The ponds were dug manually, starting from the shallow end set at the point to which floods normally reached. The slopes of the pond walls were set at a minimum of 120° in order to prevent collapse (Figure 1.4). Pond depth was adduced from the topographical maps taking into consideration the hydrological conditions and requirements for Fingerponds (natural flooding of pond system and fluctuations of the lake level). Soil from the pond was heaped to form the raised beds. Surplus excavated earth was spread outside the pond area towards the landward side. To reinforce weak areas, clay was dug from the surrounding wetland and used to line the walls of the pond.

Labour requirements

The ponds were dug by nine or more able-bodied men. The process was faster in Gaba, taking 10 days per pond on average while in Walukuba it took 21 days per pond. In Walukuba heavy rains caused collapse of the walls that necessitated removal of more earth and further slanting of the walls. The slopes and edges were then planted with grass to stabilize the construction.

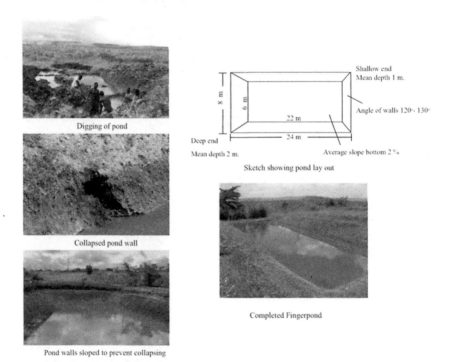

Figure 1.4 Fingerpond construction and lay out. Photographs by R.C. Kaggwa.

Wetland characterization

The characteristics of the dominant vegetation and the associated flora for the two wetlands were determined in February, 2002 at the onset of the research study. Two transects were made in each wetland in the form of belts (width 1 m) from the peripheral of the Fingerpond area up to the open water of the lake (wetland-lake interface). Cutting of transects and identification of plants were done as described in Kaggwa *et al.* (2005). The standing biomass for the two dominant vegetation types was determined using the harvest method (Kaggwa *et al.*, 2001).

Table 1.1. Plant species composition in Gaba and Walukuba wetlands

List of plant species associated with the dominant plant types

Gaba
Cyperus papyrus L. in association with

(a) *Phragmites mauritianus* (Kunth.) and sections of *Pennisetum purpureum K. Schumach*
(b) Climbers: *Ipomoea wightii* (Wall.) Choisy, *Hewittia sublobata* (L.f.) Kuntze, *Cayratia ibuensis* (Hook. F) Susseng & Susseng, *Commelina benghalensis* L. and *Melanthera scandens* (Schumach & Thonn.) Roberty.
(c) Ferns and herbaceous species: *Thelypteris confluens* (Thunb.) Morton, *Ludwiga abyssinica* A. Rich, *Zehneria minutiflora* (Cogn.) C. Jeffrey., *Polygonum pulchrum* Blume., *Pycnostachys coerulea* Hook, *Impatiens* sp.

*Phragmites mauritianus (*Kunth). found mainly on the landward side of the wetland in association with *Cyperus papyrus* L., *Hibiscus diversifolius* Jacq. and some shrubs: *Ssebania sesban* spp., *Triumfetta macrophylla* Schumann., *Sida rhombifolia, Abutilon longicupse, Crassocephalum picridifolium* (DC) S. Moore and *Alchomea cordifolia.*

Walukuba
Cyperus papyrus L. in association with

(a) *Phragmites mauritianus* (Kunth.) and scattered patches of *Typha domingensis*. Pers.
(b) Climbers: *Ipomoea wightii* (Wall.) Choisy, *Stephania abyssinica* (Quart-Dillion & A.Rich.) Walp., *Cayratia ibuensis* (Hook. F.) Susseng & Susseng and *Melanthera scandens* (Schumach & Thonn.) Roberty.
(c) Ferns and herbaceous species: *Thelypteris confluens* (Thunb.) Morton, *Ludwiga abyssinica* (A. Rich), *Zehneria minutiflora* (Cogn.) C. Jeffrey., *Polygonum pulchrum* Blume.
(d) Grasses: *Leersia hexandra* Sw.

*Phragmites mauritianus (*Kunth.) found mainly on the landward side of the wetland in association with *Cyperus papyrus* L., *Cyperus* sp., *Hibiscus diversifolius* Jacq. and some shrubs: *Ssebania sesban* spp., *Triumfetta macrophylla* Schumann.

Trees for both locations: *Acacia seyal spp.*

The two wetlands were dominated by a thick cover of *Cyperus papyrus* L. and designated with patches of *Phragmites mauritianus* (Kunth). In Gaba, *Cladium mariscus* Pohl ssp. *jamaicense* (Crantz) was also abundant. The *Cyperus papyrus* and *Phragmites mauritianus* stands were in association with various plant types (Table 1.1). *Cyperus papyrus* in both wetlands were more than 5 m tall with a biomass of over 3 kg dry weight m^{-2} while for *Phragmites* a height of 4.9 m was attained with a biomass of over 1.6 kg dry weight m^{-2}. Mean aerial productivities varied for the two vegetation types. *Phragmites* attained a maximum of 4.2 g m^{-2} d^{-1} compared to 6.0 g m^{-2} d^{-1} for papyrus.

Research strategy and aim of this thesis

The main aim of this thesis is to examine the importance of organic manure applications in enhancing nutrient levels, phytoplankton and periphyton productivities and ultimately fish production in seasonal wetland fish ponds, 'Fingerponds'. To illustrate the functions described in the preceding paragraphs, a conceptual diagram outlining the structure of the research is shown (Figure 1.5). The main flows and interactions in the pond system are described by answers to the following questions:

1. What are the key factors determining the aquaculture potential of the wetland ecotone?
2. What changes occur in pond water quality in Fingerponds after application of manure and what are the impacts on phytoplankton assemblages and productivity?
3. What relationships exist between the manure application levels, nutrient concentrations and phytoplankton primary productivity for optimization of fish production?
4. Are artificial substrates for periphyton development viable in Fingerpond systems and how do they affect pond water quality?

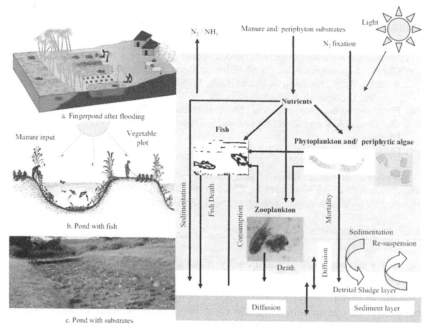

Figure 1.5. Conceptual framework of a functional Fingerpond. a. adapted from Denny *et al*., 2006, b. modified from Denny, 1991, c. A Fingerpond in Walukuba with artificial substrates installed. Photograph by R.C. Kaggwa

5. What are the key factors driving fish production in Fingerpond systems?
6. What are the effects of input factors (nutrients) and pond conditions (light, water levels, temperature) in determining food availability (phytoplankton and periphyton) for fish growth in an organically fertilized Fingerponds?

In answering these questions this thesis attempts to provide a basis for the management of seasonal wetland fishponds under the concept of wise use of wetlands.

Research scope

This dissertation focuses on the processes and management of the Fingerpond system following enhancement of primary productivity and fish production: it examines the underlying factors affecting both processes and management, qualitatively and quantitatively, by using field data and available literature. The results are combined into a dynamic pond model that gives insight into the functioning of the pond system based on pond inputs/conditions and fish growth. Finally management options are given on how best to maintain /optimize fish production in these systems.

Objectives of this thesis

The specific objectives of this research were:-
1. To examine key factors determining the aquaculture potential of the ecotone for aquaculture, such as soil conditions, flooding, natural stocking, and water quality in the ecotone and ponds in order to obtain baseline information for a better understanding and management of the Fingerponds environment.
2. To determine the relationships between manure applications, nutrient levels and phytoplankton production for optimization of fish production.
3. To determine the viability of introduced substrates for periphyton development as a food resource in Fingerponds.
4. To assess the fish production potential of organically manured ponds in the presence of artificial substrates and determine the key factors driving fish production in Fingerpond systems.
5. To develop a dynamic model that elucidates the relationship between the key drivers; pond inputs, pond conditions, natural food production and fish growth.

Outline of thesis

This dissertation is divided into seven chapters. Chapter 1 (this chapter) provides insight into the current challenges Africa is facing with respect to provision of food security as well as the linkage between natural wetlands and fisheries. It describes the wetlands used in the research and highlights the construction of Fingerponds, scope and objectives of the study.

In Chapter 2, an assessment was made of the wetland ecotone to establish the conduciveness of the environment and its potential for fish culture. The use of organic manure as a food source and the effects on pond water quality and primary productivity were examined in Chapter 3. The limitations to the effectiveness of organic manure in such pond systems were highlighted. As an additional drive to provide alternative low cost food sources for the fish, the potential use of artificial substrates for development of periphyton was examined in Chapter 4.

Chapter 5 focused on fish production and the effects of enhanced nutrient levels and artificial substrates. In Chapter 6, the information gathered, together with data from literature, was integrated into a dynamic pond model that established links and interactions between fish growth, pond inputs, water quality, detritus and sediment.

Chapter 7 concludes the dissertation with a general discussion that concentrates on critical limitations that resulted in low fish production, identifies gaps in the current knowledge and gives useful recommendations for management of Fingerpond systems as well as highlighting areas for future research.

References

Azim, M.E., Verdegem, M.C.J., van Dam, A.A. and Beveridge, M.C.M. 2005. *Periphyton. Ecology, exploitation and management.* CABI Publishing. 319 pp.

Balirwa, J.S. 1998. *Lake Victoria wetlands and the ecology of the Nile Tilapia Oreochromis niloticus Linné.* A.A. Balkema Publishers, Rotterdam, The Netherlands. 247 pp.

Benson, T. 2004. Africa's food and nutrition security situation. Where are we and how do we get there. Paper commissioned for the International Food Policy Research Institute (IFPRI) 2020 Africa Conference on "Assessing food and nutrition security in Africa by 2020: Prioritizing actions, strengthening actors and facilitating partnerships'. Held in Kampala, Uganda April 1-3, 2004. *International Food Policy Research Institute*, Washington DC, USA. 86 pp.

Brummett, R.E. and Noble, R. 1995. Aquaculture for African Smallholders. *ICLARM* Tech. Rep. 46. 69 pp.

Bugenyi, F.W.B. 1991. Ecotones in a changing environment: Management of adjacent wetlands for fisheries production in the tropics. *Verh. Internat. Verein. Limnol.* 24: 2547-2551.

Coche, A.G., Haight, B.A., Vincke, M.M.J. 1994. Aquaculture development and research in Sub-Saharan Africa. Synthesis of natural reviews and indicative action plan for research. CIFA, Technical Paper No. 23. Rome, *FAO.* 151 pp.

Colman, J.A. and Edwards, P. 1987. Feeding pathways and environmental constraints in waste-fed aquaculture: balance and optimization. In: Moriarty, D.J.W. and Pullins, R.S.V. (eds.) *Detritus and microbial ecology in Aquaculture.* ICLARM, Manila. 240 pp.

Costa-Pierce, B.A. 2002. *Ecological aquaculture: The evolution of the blue revolution.* Blackwell Publishing Ltd. 400 pp.

Delgado, C.L., Wada, M., Rosegrant, N.W., Meijer, S. and Ahmed, M. 2003. Outlook for Fish to 2020. Meeting Global Demand. A 2020 Vision for Food, Agriculture and the Environment Initiative. *International Food Policy Research Institute*, Washington, D.C. USA, World Fish Centre, Penang Malaysia. 36 pp.

Delincé, G. 1992. *The ecology of the fish pond ecosystem with special reference to Africa.* Kluwer Academic Publishers, Dordrecht. 230 pp.

Denny, P. 1985 (ed.) *The ecology and management of African wetland vegetation.* Geobotany 6. Dr. W. Junk Publishers, Dordrecht, The Netherlands. 344 pp.

Denny, P. 1991. African wetlands. In: M. Finlayson and M. Moser. *Wetlands.* International Water Fowl and Wetland Research (IWRB). Facts on File, Oxford and New York. pp. 115-148.

Denny, P. and Turyatunga, F. 1992. Uganda wetlands and their management. In: E. Maltby, P.J. Dugan and J.C. Lefeuvre (eds.). *Conservation and development: The sustainable use of wetland resources. Proceedings of the third international wetland conference.* Rennes, France, 19-23 September, 1998. IUCN, Gland, Switzerland. Xii + 219 pp.

Denny, P., Kipkemboi, J., Kaggwa, R. and Lamtane, H. 2006. The potential of fingerpond systems to increase food production from wetlands in Africa. *International Journal of Ecology and Environmental Sciences* 32(1): 41-47.

Drammeh, 2001. Illegal Unreported and Unregulated fishing in small-scale marine and inland capture fisheries. In: FAO Fisheries Report No. R666. Report of papers presented at the expert consultation on illegal unreported and unregulated fishing, Sydney, Australia. 15-19 May 2000. 309 pp.

Dugan, P., Dey, M. and Sugunan, V.V. 2004. Fisheries and water productivity in tropical river basins: enhancing food security and livelihoods by managing water for fish. In: New direction for a diverse planet. Proceedings of the 4th international crop science congress. 26 Sept-1 Oct 2004. Brisbane, Australia. 10 pp.

Egna, H.S. and Boyd, C.E. 1997. *Dynamics of pond aquaculture.* CRC Press, Boca Raton/New York. 437 pp.

FAO-SAFR, 1998. Wetland characterization and classification for sustainable agriculture development. *FAO-*Sub Regional Office for East and Southern Africa, Harare.

FAO 2000. The state of the world fisheries and aquaculture (SOFIA). *FAO.* Rome. 141 pp.

Geheb, K. and Binns, T. 1997. 'Fishing farmers' or Farming fishermen'? The quest for household income and nutritional security on the Kenyan shores of Lake Victoria. *African Affairs* 96: 73-93.

FAO, 2001. Integrated agriculture-aquaculture: a primer. *FAO/IIRR/World Fish Center.* 149 pp.

FAO, 2004a. The state of the world fisheries and aquaculture (SOFIA). *FAO.* Rome 153 pp.

FAO, 2004b. Aquaculture extension in Sub-Saharan Africa. *FAO.* Fisheries circular No. 1002. Rome. 55 pp.

Halwart, M. and van Dam, A.A. (eds.) 2006. *Integrated irrigation and aquaculture in West Africa. Concepts, practices and potential,* Rome, FAO. 196 pp.

Hishamunda, N., Thomas, M., Brown, D., Engle, C. and Jolly, C. 1998. Small-scale fish farming in Rwanda. Economic considerations. *CRSP Research Report* 98-124. 12. pp.

Holland, M.M. 1988. SCOPE/MAB Technical consultations on landscape boundaries. Report of a SCOPE/MAB Workshop on ecotones. In: F. di. Castri, A.J. Hansen and M.M. Holland (eds.). A new look at Ecotones: Emerging International Projects on Landscape Boundaries. 47-106. *Biology International,* Special Issue 17. Paris.

Jamu, D.M. and Ayinla, A. 2003. Potential for the development of aquaculture in Africa. *NAGA, World Fish Centre. Quarterly* 26 (3): 9-13.

Kaggwa, R.C., Mulalelo, C.I., Denny, P. and Okurut, T.O. 2001. The impact of alum discharges on a natural tropical wetland in Uganda. *Water Research* 35 (3): 795-807.

Kaggwa, R.C., Kansiime, F., Denny, P. and van Dam, A.A. 2005. A preliminary assessment of the aquaculture potential of two wetlands located in the northern shores of Lake Victoria, Uganda. In: J. Vymazal (eds.) *Natural and Constructed Wetlands: Nutrients, Metals and Management.* Backhuys Publishers, Leiden, The Netherlands. pp. 350-368.

Kipkemboi, J. 2006. Fingerponds: Integrated seasonal aquaculture in East African Fresh water wetland. Exploring their potential for wetland wise-use strategies. Ph.D Thesis, UNESCO-IHE Institute for Water Education, The Netherlands (in prep.).

Knud-Hansen, C.F. 1998. Pond fertilization ecological approach and practical application. *PD/A CRSP.* 125 pp.

Kudhongania, A.W. and Cordone, A.J. 1974. Batho-spatial distribution patterns and biomass estimates of the major demersal fishes in Lake Victoria. *Afr. J. Trop. Hydrobiol.* Fish 3: 15-31.

Lowe–McConnell, R.H. 1975. *Fish communities in tropical freshwaters: their distribution, ecology and evolution.* Longman Inc. New York. 337 pp.

McCartney, M.P., Masiyandima, M. and Houghton-Carr, H.A. 2005. Working wetlands: classifying wetland potential for agriculture. Research Report 90. Colombo, Sri Lanka. *International Water Management Institute* (IWMI). 35 pp.

Pechar, L. 2000. Impacts of long-term changes in fishery management on the trophic level water quality in Czech fish ponds. *Fisheries Management and Ecology.* 7: 23-31.

Prein, M., Ofori, J.K. and Lightfoot, C. (eds.) 1996. Research for the future development of aquaculture in Ghana. *ICLARM.* Conference Proceedings. 42. 94 pp.

RCS, 2004. Ramsar handbook for the wise use of wetlands. 2nd Edition. *Ramsar Convention Secretariat*, Gland, Switzerland. 25 pp.

RCS, 2006. Ramsar information paper no.1. What are wetlands. http://www.ramsar. org/about/about_infopack_1e.htm.

Toews, D. and Griffith, J.S. 1979. Empirical s of potential fish yields for the Lake Bangweulu system, Zambia, Central Africa. *Trans. Amer. Fish. Soc.* 108: 241-252.

UN, 2005. The Millennium Development Goals Progress Report. *The UN Department of Public information.* DPI/2390. 48 pp.

UNICEF, 2001. State of the World's Children (SOWC), *UNICEF*, New York, USA. 116 pp.

UNICEF, 2005. State of the World's Children (SOWC). *UNICEF*, New York, USA.

WDI, 2005. World Development Indicators 2005. *World Bank.* 440 pp.

Welcomme, R.L. 1979. Fisheries ecology of flood plain rivers. Longman, London. 317 pp.

WFC, 2005. Fish and Food security in Africa. *World Fish Centre.* Penang. Malaysia. 11 pp.

WI, 2006. Wetlands and Poverty Reduction Project Proposal. Wageningen, The Netherlands. *Wetlands International.* 35 pp.

Williams, M. 1996. The transition in the contribution of living aquatic resources to food security. Food, Agriculture or the Environment. Discussion paper 13. *International Food Policy Research Institute*, Washington, DC. 41. pp.

Witte, F., Goldschmidt, T. and Wanink, J.H. 1995. Dynamics of the haplochromine cichlid fauna and other ecological changes in the Mwanza gulf of Lake Victoria. In: T. Pitcher and P.J.B. Hart (eds.). *The input of species changes in African lakes.* Chapman and Hall, London. pp. 83-110.

Chapter 2

Fringing wetlands on the northern shores of Lake Victoria, Uganda and their potential for fish culture

Abstract

Fingerponds are earthen ponds dug at the edge of a wetland and stocked naturally during flooding. Fingerponds aim at enhancing fish production from wetlands with minimal impact on their natural functions. This study investigated the base-line conditions of the ponds and surrounding wetland prior to the formulation of management strategies. Eight 192 m^2 ponds were constructed in two papyrus wetlands on the northern shores of Lake Victoria, Uganda. The ponds were naturally stocked predominantly by *Oreochromis spp.* during the May to June 2003 rains. Between February 2002 and September 2003, factors determining the natural dynamics and functioning of the Fingerponds and their potential for fish production were examined. Key ecological processes affecting water quality in the ecotone and the ponds were the buffering capacity, and in the ponds themselves sedimentation/re-suspension. The water quality in the ponds was stable and showed little seasonal or spatial variation. Pond water pH values ranged from 5.3 to 9.4, temperatures from 20 to 30 $^{\circ}$C, dissolved oxygen from 2.0 to 17.8 mg L^{-1} and ammonium nitrogen from 0 to 3.6 mg L^{-1}. Low nutrient concentrations in the ponds resulted in low phytoplankton biomass. Zooplankton colonization of the ponds was poor resulting in low biomass. Flood events determined the type and size of fish in the ponds. Initial fish stock biomass at the onset of the 2003 grow-out period ranged between 93 and 97 kg ha^{-1}. It was deduced that with enhancement of nutrient levels, Fingerpond systems have potential and can be of real benefit as a source of protein to poor, riparian communities.

Key words: ecotone, fingerponds, potential, processes, water quality, wetland,

Publication based on Chapter 2:
Kaggwa, R.C., Kansiime, F., Denny, P. and van Dam, A.A. 2005. A preliminary assessment of the aquaculture potential of two wetlands located in the northern shores of Lake Victoria, Uganda. In: J. Vymazal (eds.) *Natural and constructed wetlands: Nutrients, metals and management*. Backhuys Publishers, Leiden, The Netherlands, pp. 350-368.

Introduction

In East Africa, water regimes of floodplains and other wetlands are mainly determined by rainfall. During flooding, inundated areas provide a habitat for aquatic organisms. Together with the release of nutrients from submersed soil, this produces a surge of primary productivity closely followed by an expansion of the biomass of aquatic organisms (Chapman *et al.*, 2004). When the flood retreats, fish either select refugia or are trapped in shallow pits within the swamp (Magoulick & Kobza, 2003). Traditional wetland aquaculture systems such as brush parks and culture-based fisheries in floodplains or wetlands utilize the productivity of wetlands for producing much-needed fish protein for riparian communities. Such traditional systems are widespread in Africa (Jamu & Ayinla, 2003) and Asia (Prein, 2002). Many attempts at developing aquaculture in Africa have focused on improving the productivity of earthen ponds through formulated feeds and hatchery-raised fingerlings. Such technology is often out of reach of resource-poor wetland communities. Furthermore, little attention has been given to the improvement of traditional aquaculture systems.

The use of wetland interface zones for increasing swamp fish production through the so-called "Fingerponds" has been proposed (Denny, 1989; Denny & Turyatunga, 1992). Fingerponds are earthen ponds dug at the edge of a wetland, filled and stocked naturally with fish during flooding. The concept relies entirely on seasonal flood events and has minimal interference with the integrity of the wetland environment. Fingerponds exploit (a) the high natural productivity of the swamp/lake interface (which provides the fish stock for the ponds); and (b) the trapping of fish in depressions following the retreat of floods after seasonal rains.

In Lake Victoria wetlands, seasonal patterns in fish abundance and diversity depend on vegetation type and zoobenthic productivity (Balirwa, 1998). It is not clear which fish species can penetrate the dense papyrus wetlands. The physico-chemical environment of the Fingerpond system within the wetland ecotone is poorly understood. The limnological and sediment characteristics of newly established ponds prior to and after natural stocking and the suitability of this environment for aquaculture also need to be established. The main objective of this study was to obtain baseline information for a better understanding and management of the Fingerponds environment. The key factors determining the aquaculture potential of the ecotone for aquaculture, such as soil conditions, flooding, natural stocking, and water quality in the ecotone and ponds, are examined. Based on this knowledge, guidelines for management and enhanced fish production in Fingerponds are proposed.

Materials and Methods

Study area and period

The study was carried out from February 2002 to September, 2003. Four Fingerponds (24 x 8 m, 1.5 m average depth) were constructed in each of two localities on the northern shores of Lake Victoria, Uganda; Gaba and Walukuba. The Gaba wetland borders the Inner Murchison bay, 13 km South East of Kampala. It is set at $0^{\circ} 15$'N, at 1143 m above sea level and covers an area of 1.43 km^2. The

Walukuba wetland borders the Napoleon Gulf in Jinja between $0°$ 22'N and $0°$ 30'N and longitude $22°10$'E and $33°$ 26'E at 1143 m above sea level and covers an area of 3 km^2. The two wetlands are dominated by *Cyperus papyrus L.* and *Phragmites mauritianus* (Kunth) (see Chapter 1). In February 2002, two 1 m wide transects were made through the wetland 20 and 70 m from the peripheral of the Fingerpond area up to the lake. Transect lengths were 190 (T-W1) and 50 m (T-W2) in Walukuba and 50 (T-W1) and 100 m (T-W2) in Gaba (Figure 2.1).

This part of Uganda receives seasonal rainfall twice a year from March to May and from September to November, averaging 1000 mm per year. The average water level in the lake normally fluctuates by 0.5 m (Mnaya & Wolsanki, 2002). During 2002, the March-May floodwaters from the lake were insufficient to fill the ponds either in Gaba or Walukuba: the ponds were still under construction. The Gaba ponds, completed at the end of May, were filled by digging a shallow inlet channel from the lake to the ponds (Figure 2.1). In 2003, the flood was higher and filled the ponds at both locations with water and fish.

The study covered three sampling periods. During Period 1 (March-November 2002; 36 weeks), data on soil, vegetation and water quality were collected in the wetlands only. Data on vegetation is presented in Kaggwa *et al.* (2005). In Period 2 (December 2002-January 2003; 8 weeks), water quality data were taken in the whole ecotone (lake, wetland and ponds). In Period 3 (May-September 2003; 12 weeks in Gaba and 17 weeks in Walukuba), water quality and plankton data were collected in the ponds after they had been flooded.

Water levels and sediment

Ground water levels in the wetlands were determined between February and March 2002 and at the onset of the rainy season in April 2003 by auguring holes at 10 m intervals along the transects. Rainfall was recorded daily at 09.00 hours using a rain gauge positioned at each site. Sediment cores were collected in March 2002 from six transect points using a hand-operated, 5-cm diameter core sampler. Pond sediment samples were collected in June 2002 (prior to filling of the ponds) and in June 2003 (after flooding). In each pond, 5-cm deep samples were taken along a 'Z' curve and pooled into three composite samples that were placed in black polyethylene bags and kept on ice and thereafter transported to the laboratory. Wet soil pH (direct, glass electrode) and conductivity (1:2 water extract) were measured immediately. Dry weights (oven dried: 60°C, 48 h) were determined, samples ground and stored at room temperature prior to further analysis. Sub-samples were analyzed for texture (hygrometer method), organic matter (loss on ignition; Klute, 1986; Okalebo *et al.*, 1993), total nitrogen and phosphorus (TN and TP; sulphuric acid-selenium method; Novozamsky *et al.*, 1984), calcium and magnesium (Ca and Mg; atomic absorption), and sodium and potassium (Na and K; flame photometry) according to standard methods (APHA, 1995).

Water quality

During Period 1, water quality in the wetlands was determined every two months along the transects at 10 m intervals. In Period 2, sampling was done fortnightly from the inshore lake zone, the wetland and the ponds. In the lake, six points along two transects set 0.5 m and 20 m from the lake edge were sampled from the surface, middle and bottom (T-L1 and T-L2; Figure 2.1). In the wetlands, the transects were sampled as in Period 1 while in the ponds, subsurface samples were taken at the shallow, middle and deep ends. During Period 3, sub-surface samples from the ponds were taken fortnightly. All sampling was done between 10.00 and 14.00 hours. *In situ* measurements were made for pH, conductivity, temperature and dissolved oxygen (DO) using handheld meters (MODEL 340i, WTW, Weilheim, Germany). All physico-chemical analyses for ammonium nitrogen (NH_4-N) (APHA, 1992), total suspended solids (TSS), turbidity, nitrate nitrogen (NO_3-N), total nitrogen (TN), orthophosphate (PO_4-P), total phosphorus (TP), alkalinity, chloride and biochemical oxygen demand (BOD) were carried out according to standard methods (APHA, 1995).

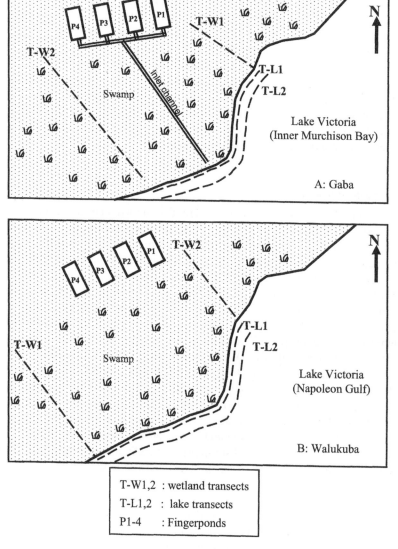

Figure 2.1. Schematic layout of Fingerponds in Gaba and Walukuba.

Phytoplankton, zooplankton and fish
Phytoplankton samples of known volumes were collected from the lake, wetland and ponds with a dip sampler and preserved with 1% Lugol's solution. Samples were taken fortnightly in Period 2 and monthly in Period 3 between 10.00 and 14.00 hours. In the ponds, integrated samples were taken from the deep end. Sedimentation and enumeration of phytoplankton was done by inverted microscope (modified Utermöl method; Nauwerck, 1963). Identification was done up to genus level. Biomass was determined as chlorophyll a (Chl a) based on the acetone extraction method (Wetzel & Likens, 1991).

Zooplankton samples were collected from the lake (vertical hauls) and ponds (horizontal hauls) using an Apstein net (80 μm mesh). Samples were taken fortnightly in Period 2 between 10.00 and 14.00 hours. For the ponds, samples from the shallow, middle and deep ends were pooled. In the wetland, known volumes were collected from pools of water adjacent to the transects. Samples for taxonomic identification up to genus level were preserved with formalin solution to a final concentration of 4%. Zooplankton densities and dry weight biomass were found as described by Duncan (1975) and Fernando (2002).

Fish samples in the wetland were collected from pools and transported alive in water containers to the laboratory for identification. Fish samples were collected during Period 3 in the months of April and May. In each pond, three successive catches were made with a seine net (12 x 2 m, mesh size 6.5 mm) and individual fresh weights and total lengths were determined. Fish were identified according to Greenwood (1966). Total population biomass was estimated by extrapolation of the catches. Fish were preserved in formalin (10 % for fish >10 cm, 5 % for fish < 10 cm) and gut analysis carried out in the laboratory using the percentage occurrence method as described by Hyslop (1980) and Balirwa (1998).

Statistical analysis
Statistical analysis was performed using SPSS 11.0 (SPSS Inc, Chicago, Illinois, USA). Differences in means were tested with the independent t-test and homogeneity of variance with Levene's test. Differences were considered significant at a level of $P=0.05$. Coefficients of variance (C.V) were used to show variability of mean values. Water quality data of the wetland, the ecotone and the ponds in Periods 1, 2 and 3 were analyzed using factor analysis (Milstein, 1993). Three datasets were created, with 215, 209 and 276 cases, respectively. Apart from water quality data, some extra variables were introduced such as the rainfall in the week prior to sampling (in mm) and the pond volume at the time of sampling (m^3). Principal components were extracted using Varimax rotation. Separate analyses were done for Gaba and Walukuba. Only factors with an eigenvalue of 1 or more were retained, resulting in four to six factors for each dataset. The factor loadings were interpreted using sign and relative size of the coefficient as an indication of weight placed upon each variable. Factor loadings with values larger than 0.50 were considered significant.

Results

Water levels within the wetland prior to flooding

Monthly total rainfall over the three experimental periods followed the typical wet (March-June, September-November) and dry season pattern (Figure 2.2). In 2002, little variation in water levels was noted in the shorter transects at the onset of the rains (February/March). In the longer transect in Gaba (T-W2), water rose above the ground to about 100 cm around the lake edge in the sixth week of monitoring but did not reach and flood the ponds. The water was directed into an inlet channel 0.5 m wide to feed the ponds (Figure 2.1). The ponds in Walukuba were not filled. In 2003 (April/May), water levels in the longer transects rose to about 30 cm in Gaba and 10 cm in Walukuba and reached the ponds. These levels were sufficient to fill all ponds though the ponds were not fully inundated.

Wetland sediment and pond soil characteristics

Results of the sediment analysis are presented in Table 2.1. In the wetlands, pH values were near to neutral (6.7-6.9). Na, Ca and Mg concentrations did not vary significantly between the two wetlands whilst in Gaba the K values were significantly lower (t test, $p<0.05$). The sediment in Gaba had a sandy loam texture with significantly more sand (t test, $p<0.05$) compared to a silty loam texture in Walukuba. The clay and organic matter contents in the two locations were similar, as were TP and TN concentrations although for TN differences were found between transects: mean $\pm$ standard error of the mean (SE) 0.72 ± 0.69 mg L^{-1} (T-W1) and 0.24 ± 0.09 mg L^{-1} (TW-2). In the ponds, soils were alkaline with pH values of 7.8-8.7. The organic matter content in the top 5 cm was higher in Walukuba (10-12 %) whereas pond soils in Gaba had higher clay contents. Generally nutrient concentrations were higher in Walukuba than in Gaba and showed notable differences between 2002 and 2003 (Table 2.1). In 2002, the elements concentration was higher in all ponds whilst the clay content in the Walukuba ponds was higher in 2003 at 43.7 % compared to 25.8 % in the previous year but demonstrated high variability with a coefficient of variance (C.V) of 122 %.

Water quality in the wetland, (March-November, 2002), Period 1

The water quality data for the wetlands are given in Table 2.2. Mean water temperatures were lower in Gaba (22.2-22.5 °C) than in Walukuba (24.0-24.7 °C) and pH values ranged from 6.4 to 7.0 in Gaba and from 5.9 to 7.7 in Walukuba with low C.V.s for both parameters. The other variables had high C.V. above 100 % which implied temporal variation. Ammonium-nitrogen (NH_4-N) concentrations varied from 0.06 to 11.45 mg L^{-1} (Gaba) and 0.05 to 36.75 mg L^{-1} (Walukuba) whilst orthophosphate (PO_4-P) did not show much variation between locations; 0.002-1.050 mg L^{-1} (Gaba) and 0.003-1.965 mg L^{-1} (Walukuba). Higher total phosphorus (TP) concentrations were found in T-W1, Walukuba with values going up to 10 mg L^{-1} while total nitrogen (TN) and nitrate nitrogen (NO_3-N) showed no particular pattern in both locations (NO_3-N: 0-0.93 mg L^{-1}, TN: 0.06-133 mg L^{-1}). Chloride concentrations ranged from 15 to 75 mg L^{-1} (Gaba) and 2 to 266 mg L^{-1} (Walukuba) with high values noted in Walukuba in March.

Dissolved oxygen (DO) concentrations were generally low but rose to values exceeding 2 mg L^{-1} towards the lake (Figure 2.3). In both locations, alkalinity ranged from 0 to about 1000 mg $CaCO_3$ L^{-1}, with lowest values occurring close to the lake. EC followed a similar pattern but with a wider range from 28 to 2150 μS cm^{-1} in Walukuba compared to 89 to 1460 μS cm^{-1} in Gaba. BOD_5, NH_4-N, total suspended solids (TSS) and turbidity changed considerably with time (season) with the highest values occurring in months that received most rainfall (Figure 2.2). TN and TP decreased with distance (Figure 2.3) but with high temporal variation particularly at the lake-wetland interface.

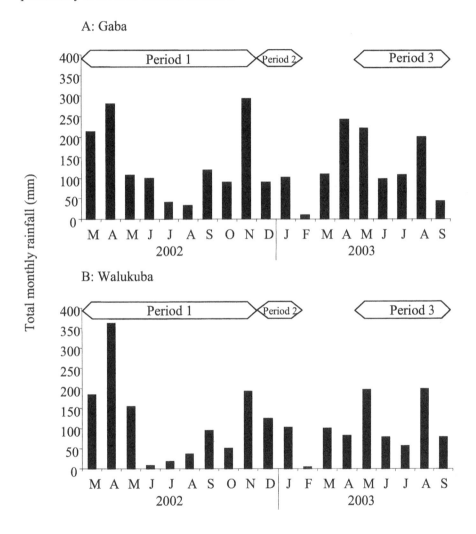

Figure 2.2. Rainfall pattern during study period (Periods 1, 2 and 3).

Table 2.1. Wetland and pond sediment analysis, Periods 1 (March and June 2002) and 3 (June 2003), Gaba and Walukuba

	Gaba						Walukuba					
	Wetland transect		Ponds				Wetland transect		Ponds			
	Pre flooding, Mar-02		Pre flooding, Jun-02		Post flooding, Jun-03		Pre flooding, Mar-02		Pre flooding, Jun-02		Post flooding, Jun-03	
	Mean	C.V.	Mean	C.V.	Mean	C.V.	Mean	C.V.	Mean	C.V.	Mean	C.V.
pH	6.7	5.6	7.9	3.9	8.3	2.2	6.9	7.3	8.7	3.0	8.5	2.7
EC (μS cm^{-1})	nd		347	49.9	269	21.8	nd		632	88.0	674	13.6
Eh (mV)	nd		−69	23.1	−70	13.2	nd		−118	12.4	−92	13.8
K (mg g^{-1})	0.105 *	20.9	0.348	31.2	0.01	138.5	0.208	26.4	0.115	29.8	0.029	13.2
Na (mg g^{-1})	0.013	18.5	0.02	39.9	0.016	50.9	0.018	20.4	0.626	43.2	0.094	36.7
Ca (mg g^{-1})	2.253	13.0	0.95	27.0	0.198	118.5	2.443	14.9	4.012	10.5	0.53	44.3
Mg (mg g^{-1})	0.338	8.7	0.233	24.4	0.048	41.9	0.386	10.6	1.749	2.4	0.144	37.4
% Sand	60.2	58.7	38.5	98.6	40.5	54.0	43.5 *	25.9	28.8	21.9	41.2	32.7
% Clay	23.7	23.6	41.9	64.1	43.7	50.2	25.8	16.1	29	31.4	43.7	121.6
% Silt	16.2 *	9.7	21	9.9	15.8	8.7	30.7	14.4	42	12.7	15.2	5.3
% O.M.	3.8	26.5	8.2	14.7	7.4	14.0	4.6	24.1	12	12.5	10.2	6.2
% N	0.43	21.6	0.04	12.1	0.01	7.0	0.56	20.5	0.28	10.0	0.03	5.8
% P	0.07	8.5	0.01	23.3	0.02	24.4	0.05	23.2	0.08	9.9	0.06	22.8
% C	2.23	-	4.76	-	4.29	-	2.65	-	6.96	-	5.92	-

Values represent means and coefficients of variance (C.V.s) for one sampling date for six sampling points in the transects and three composite samples in the ponds.
* indicates values significantly lower than the other site. nd = not determined.

Table 2.2. Water quality analysis in wetland transects (T-W) in Gaba and Walukuba, Period 1.

	Gaba				Walukuba			
	T-W 1 (n = 30)		T-W 2 (n = 55)		T-W 1 (n = 100)		T-W 2 (n = 35)	
	Mean	C.V.	Mean	C.V.	Mean	C.V.	Mean	C.V.
pH	6.6	6	6.7	5	6.5	9	7.0	6
EC (μS cm^{-1})	268	105	277	62	391	103	734	86
Temperature (°C)	22.5	8	22.2	6	24.0	8	24.7	5
Alkalinity (as CaCO$_3$)	153	135	127	95	178	125	286	80
DO (mg L^{-1})	0.9	55	1.0	100	0.7	152	0.6	55
Turbidity (NTU)	664	205	178	169	198	129	71	148
TSS (mg L^{-1})	510	307	81	159	350	150	192	162
Cl$^-$ (mg L^{-1})	34.5	106	26.2	58	32.4	101	32.0	128
NH$_4$-N (mg L^{-1})	2.559	134	1.600	116	4.183	149	4.143	164
NO$_3$-N (mg L^{-1})	0.033	117	0.019	99	0.072	184	0.046	191
PO$_4$-P (mg L^{-1})	0.221	131	0.099	184	0.179	158	0.204	166
TP (mg L^{-1})	1.045	120	0.305	136	1.010	141	0.729	102
TN (mg L^{-1})	12.30	212	8.60	186	6.80	171	11.60	262
BOD$_5$ (mg L^{-1})	4.5	112	5.9	132	9.2	149	7.4	111

Values represent means of 5 sampling dates and 6 and 11 sampling points in Gaba and 20 and 7 sampling points in Walukuba for T-W1 and T-W2 respectively and coefficients of variance (C.V).

With factor analysis, five factors explained a total of 72% and 63% of the variation in Period 1 in Gaba and Walukuba, respectively (Table 2.3). In both locations, F1 (20.2 and 16.5%, respectively) was correlated with EC and alkalinity. In Gaba, F1 was further related to chloride, TP and NH_4-N whereas in Walukuba it was related to distance from the lake edge, pH and NO_3-N. F2 (both locations about 14%) was related to TSS. In Gaba, F2 was also related to NO_3-N while in Walukuba it was related to NH_4-N. F3 (13% and 11%, respectively) was related to temperature, negatively in Gaba and positively in Walukuba. In Gaba, F3 was furthermore related to distance to the lake and negatively to DO. In Walukuba, F3 was positively related to both PO_4-P and TP. The fourth factor (F4, 13% and 11%) was related to TN in both locations. In Gaba, it was furthermore related to pH (negative) and PO_4-P, whereas in Walukuba it was related to BOD. The fifth factor explained another 12% and 10%, respectively and was related to BOD, turbidity and NH_4-N in Gaba and to pH, DO and chloride in Walukuba.

Water quality and plankton in the ecotone, (December 2002-January 2003), Period 2

Water quality in the ecotone zones is summarized in Table 2.4. Differences in lake water quality among the sites were negligible, except for zooplankton abundance and chlorophyll a (Chl a) which were higher in Gaba. In the wetland zone, zooplankton abundance, TSS, NO_3-N, PO_4-P and Chl a were higher in Gaba; all the other parameters were higher in Walukuba. Temperature was higher in Walukuba (mean of 24.0 °C) than in Gaba (mean 21.8 °C). In the ponds turbidity, TSS, NH_4-N, PO_4-P and zooplankton abundance were higher in Gaba. Temperature was higher in Walukuba (mean 25.1 °C) than in Gaba (mean 22.5 °C) as well as EC, DO, alkalinity, TN, BOD and Chl a. The only water quality parameter which was similar at both locations in all zones of the ecotone was pH.

Within the ecotone, mean BOD_5, TP and TN were relatively constant regardless of distance to the lake. PO_4-P, NH_4-N and TSS were generally higher in the wetland than in the lake or ponds, while Chl a, DO, pH and temperature were generally lower. Turbidity, alkalinity, EC, NO_3-N and NH_4-N increased with increasing distance from the edge of the lake whilst the zooplankton abundance decreased. In Walukuba, the ponds had lower turbidity than the wetland whereas in Gaba it was the reverse. Within each zone, pH and temperature had the least variation shown by low C.V.s around 10 or lower. All dissolved nutrients had very high variability (C.V. greater than 100).

In Gaba, factor analysis resulted in five factors explaining 76 % of the variation (Table 2.5). F1 (25.8%) was related strongly to distance to the lake, EC, alkalinity (all positive) and to zooplankton abundance and temperature (negative). F2 (15.1%) had strong correlations with PO_4-P, turbidity, TSS, and NH_4-N, while F3 (13.6%) was related to NO_3-N, BOD and TP. F4 (13.4%) was related mostly to DO and pH. F5 (8.7%) was related to the amount of rain in the 7 days prior to sampling.

In Walukuba, six factors were extracted explaining 75.9% of the variation. The first two factors resembled F1 and F2 in Gaba. F1 (18.8%) was related positively to EC, alkalinity and distance from the lake and negatively to zooplankton abundance. F2 (14.3%) was related to turbidity, NH_4-N and TSS. F3 (12.8%) was strongly related to rainfall (like F5 in Gaba) and BOD and less strongly to temperature. F4 (11%) was related to pH and Chl a and F5 (9.6%) negatively to TN and positively to DO and F6 (9.5%) was strongly related to both TP and PO_4-P.

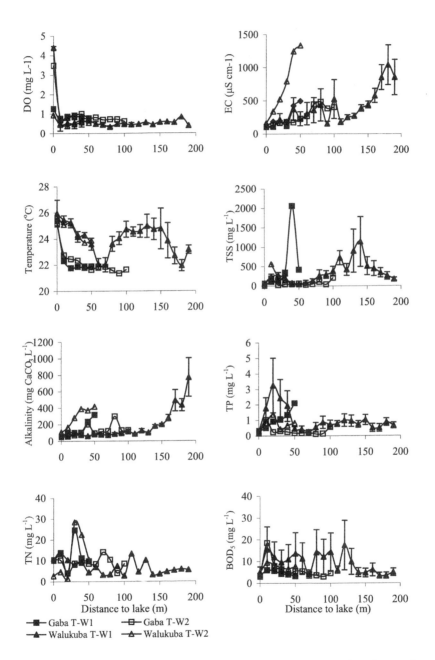

Figure 2.3. Changes in water quality parameters along transects, Period 1. Plots are for dissolved oxygen (DO), temperature (T), alkalinity, total nitrogen (TN), conductivity (EC), total suspended solids (TSS), total phosphorus (TP) and biochemical oxygen demand (BOD) for means of samples taken every two months for five months starting in March 2002 to November 2002. The bars on TW-1 Walukuba, indicate standard error of the mean and show temporal variation.

Table 2.3. Factor analysis results wetland water quality, Period 1

Gaba (N = 85)			Factors		
Variable name	F_1	F_2	F_3	F_4	F_5
Alkalinity: total (mg $CaCO_3$ L^{-1})	**0.848**	0.112	0.196	0.001	-0.126
Cl^- (mg L^{-1})	**0.827**	-0.115	0.024	0.047	0.317
EC (μS cm^{-1})	**0.737**	0.205	0.416	-0.187	-0.157
TP (mg L^{-1})	**0.690**	0.142	-0.079	0.426	0.040
TSS (mg L^{-1})	0.150	**0.907**	0.006	0.078	0.033
NO_3-N (mg L^{-1})	0.009	**0.888**	0.082	0.084	0.122
Temperature (^{o}C)	-0.175	-0.030	**-0.790**	-0.330	0.007
Distance from lake shore (m)	0.132	0.009	**0.769**	-0.366	-0.217
DO (mg L^{-1})	-0.012	-0.048	**-0.739**	-0.143	-0.257
PO_4-P (mg L^{-1})	0.133	-0.112	0.087	**0.788**	-0.124
pH	0.277	-0.143	-0.027	**-0.735**	-0.086
TN (mg L^{-1})	0.275	0.195	0.055	**0.559**	-0.148
BOD_5 (mg L^{-1})	-0.152	0.005	-0.008	0.026	**0.816**
Turbidity (NTU)	0.187	0.135	0.068	-0.178	**0.678**
NH_4-N (mg L^{-1})	**0.552**	0.486	-0.009	-0.036	**0.552**
% Variance explained	20.17	13.53	13.35	13.35	11.64

Walukuba (N = 130)	Factors				
Variable name	F_1	F_2	F_3	F_4	F_5
Alkalinity: total (mg $CaCO_3$ L^{-1})	**0.844**	0.029	0.012	0.107	0.098
pH	**0.700**	-0.171	-0.090	0.162	**-0.512**
NO_3-N (mg L^{-1})	**0.698**	0.124	0.123	-0.159	0.005
EC (μS cm^{-1})	**0.598**	-0.153	-0.210	-0.125	0.029
Distance from lake shore (m)	**0.542**	0.393	-0.265	-0.139	0.314
TSS (mg L^{-1})	-0.051	**0.906**	-0.005	0.030	0.009
NH_4-N (mg L^{-1})	-0.019	**0.760**	0.002	0.318	0.009
Turbidity (NTU)	0.065	**0.612**	0.147	-0.149	0.151
TP (mg L^{-1})	-0.066	0.237	**0.771**	-0.173	0.105
PO_4-P (mg L^{-1})	0.115	-0.066	**0.671**	0.334	0.360
Temperature (^{o}C)	-0.238	-0.043	**0.583**	0.207	-0.238
TN (mg L^{-1})	0.052	-0.013	0.237	**0.779**	0.007
BOD_5 (mg L^{-1})	-0.191	0.116	-0.069	**0.777**	0.096
DO (mg L^{-1})	0.027	-0.149	0.080	-0.006	**-0.813**
Cl^- (mg L^{-1})	0.127	-0.029	0.307	0.197	**0.525**
% Variance explained	16.53	13.97	11.30	11.05	10.25

F denotes factors, N = number of cases. Loadings in bolded letters are greater than 0.500

Table 2.4. Water quality analysis ecotone (lake, wetland and ponds), Period 2.

| | Gaba | | | | | | Walukuba | | | | | |
| | Lake (n = 24) | | Wetland (n = 28) | | Ponds (n = 48) | | Lake (n = 24) | | Wetland (n = 40) | | Ponds (n = 45) | |
	Mean	C.V.	Mean	C.V.	Mean	C.V.	Mean	C.V.	Mean	C.V.	Mean	C.V.
pH	8.5	8	7.1	4	8.6	4	8.7	6	7.1	7	8.7	10
Temperature (°C)	26.9	4	21.8	8	22.5	5	27.5	5	24.0	7	25.1	7
EC (μS cm^{-1})	106	1	233	39	541	24	99	5	553	106	1283	13
DO (mg L^{-1})	7.4	22	2.2	56	5.1	35	7.7	54	4.7	117	8.5	49
Turbidity (NTU)	11	70	82	153	277	71	13	126	164	281	60	66
TSS (mg L^{-1})	20	88	205	284	185	84	25	127	155	181	80	61
Alkalinity (as CaCO$_3$)	45	13	111	45	258	29	43	9	287	105	533	8
NO$_3$-N (mg L^{-1})	0.025	164	0.178	486	0.306	140	0.030	111	0.024	216	0.254	176
NH$_4$-N (mg L^{-1})	0.265	87	0.899	271	0.943	97	0.116	138	1.295	194	0.377	88
PO$_4$-P (mg L^{-1})	0.053	90	0.212	118	0.189	126	0.054	67	0.171	73	0.082	191
TN (mg L^{-1})	5.6	50	8.0	67	7.0	52	8.4	74	11.6	85	9.1	79
TP (mg L^{-1})	0.408	77	0.902	116	0.547	75	0.362	223	1.021	70	0.618	56
BOD (mg L^{-1})	4.5	40	5.9	63	4.8	48	5.5	69	6.0	72	8.7	50
Chl a (μg L^{-1})	186.4	86	55.0	231	51.6	107	136.4	76	45.7	70	116.1	77
Zooplankton abundance (ind. L^{-1})	1251	42	451	60	209	57	881	60	106	36	130	39

Values represent means of four sampling dates and six sampling points in the lake, seven and ten in Gaba and Walukuba wetlands respectively and three in each pond as well as coefficients of variance.

Water quality in the ponds, (May-September, 2003), Period 3

The quality of the pond water for Period 3 showed similarities between locations for some of the variables (Table 2.6). pH values ranged from 5.3-9.4 while temperature fell within the 20-30°C band with higher values noted in Walukuba. DO concentrations showed similarities in diel and temporal variation with values mostly greater than 2 mg L^{-1}. In Walukuba EC values ranged from 800-1855 μS cm^{-1} compared with 243-700 μS cm^{-1} in Gaba and alkalinity exceeded 800 mg $CaCO_3$ L^{-1} compared with less than 300 mg $CaCO_3$ L^{-1} in Gaba. Nutrients were generally low with NO_3-N < 0.1 mg L^{-1} and PO_4-P < 0.05 mg L^{-1}. TN and TP were high in the first season but gradually reduced. Gaba pond waters had high turbidities initially with values greater than 600 NTU but they reduced with time. Chl *a* in the ponds was relatively low with values for Gaba and Walukuba ranging from 4.0-183.3 μg L^{-1} and 10.7-347.5 μg L^{-1} respectively.

In Gaba, four factors were extracted and explained 69.8% of the variation (Table 2.7). F1 (32.9%) was related with pH, secchi depth, alkalinity, DO and EC (all positively) and negatively with turbidity and suspended solids. The second factor (F2, 13.3%) was related with temperature and NH_4-N nitrogen and to some extent with TN. F3 (12.3%) was related to NO_3-N and chlorophyll *a* (both positively) and negatively to pond volume while F4 was related negatively to the phosphorus variables (PO_4-P and TP).

In Walukuba, six factors explained 70.9% of the total variation. F1 (17.1%) was related to EC and alkalinity (like in Gaba), while F2 (14.1%) was related to the NO_3-N and TN and secchi depth. F3 (12.2%) was related to suspended solids and turbidity and F4 (10.0%) to DO and temperature. F5 (10.0%) was related to pH and NH_4-N and F6 (7.5%) to Chl *a* and TP.

Plankton community, Period 2 and Period 3

Phytoplankton phyla and genera type in the three zones were similar for the two locations but with noticeable differences in diversity. The number of genera found in the lake, wetland and ponds were 19, 8 and 18 for Gaba and 20, 11 and 17 for Walukuba, respectively (Table 2.8). Phytoplankton composition in the ponds was similar in both locations with six phyla: Cyanobacteria, Cryptophyta, Bacillariophyta, Dinophyta, Euglenophyta and Chlorophyta (Figure 2.4). The ponds had 35 and 32 genera in Gaba and Walukuba, respectively. The most abundant genera in Gaba and Walukuba were *Aphanocapsa* (4.81 x 10^8 cells L^{-1}) and *Ankistrodesmus* (1.2 x 10^9 cells L^{-1}).

A comparison of the percentage contribution of each phylum to the abundance and biomass over the four months showed that Walukuba had a higher abundance of Chlorophyta compared to Gaba which had more Cyanobacteria (Figure 2.4).

Table 2.5. Factor analysis results for ecotone water quality (lake, wetland and ponds), Period 2.

Gaba (N=100)	Factors				
Variable name	F1	F2	F3	F4	F5
Distance from lake shore (m)	**0.955**	0.051	-0.021	0.038	-0.001
EC ($\mu S\ cm^{-1}$)	**0.911**	0.192	0.005	0.137	-0.138
Alkalinity (as $CaCO_3$)	**0.900**	0.169	-0.018	0.114	-0.068
Zoo abundance (ind. L^{-1})	**-0.823**	0.005	0.011	0.263	-0.182
Temperature (^{o}C)	**-0.608**	-0.267	-0.170	**0.537**	-0.302
PO_4-P (mg L^{-1})	0.025	**0.827**	0.180	-0.270	0.019
Turbidity (NTU)	0.491	**0.718**	0.243	0.138	-0.033
TSS (mg L^{-1})	0.403	**0.648**	0.252	0.212	-0.238
NH_4-N (mg L^{-1})	0.046	**0.598**	-0.215	-0.164	0.303
NO_3-N (mg L^{-1})	0.160	0.225	**0.836**	0.043	0.020
BOD_5 (mg L^{-1})	-0.049	-0.246	**0.770**	-0.084	0.269
TP (mg L^{-1})	-0.066	0.325	**0.699**	-0.206	0.018
DO (mg L^{-1})	-0.204	-0.095	-0.150	**0.874**	0.081
pH	0.272	-0.087	-0.021	**0.869**	0.018
Rainfall prior 7 days (mm)	0.052	0.008	0.275	-0.005	**0.891**
TN (mg L^{-1})	-0.070	0.466	0.431	-0.215	-0.315
Chl a ($\mu g\ L^{-1}$)	-0.469	0.069	-0.131	0.321	0.463
% Variance Explained	25.76	15.13	13.60	13.37	8.73

Walukuba (N=109)	Factors					
Variable name	F1	F2	F3	F4	F5	F6
EC ($\mu S\ cm^{-1}$)	**0.927**	0.030	0.056	0.117	-0.071	-0.113
Alkalinity (as $CaCO_3$)	**0.915**	-0.012	0.025	0.079	0.014	0.051
Distance from lake shore (m)	**0.832**	0.231	0.038	-0.111	0.143	0.076
Zoo abundance (ind. L^{-1})	**-0.608**	-0.119	-0.113	0.521	0.113	-0.101
Turbidity (NTU)	0.126	**0.907**	-0.011	-0.073	-0.005	0.007
TSS (mg L^{-1})	0.265	**0.805**	-0.222	0.040	0.035	0.098
NH_4-N (mg L^{-1})	-0.034	**0.755**	0.094	-0.167	-0.147	0.221
Rainfall prior 7 days (mm)	-0.057	-0.079	**0.886**	-0.030	-0.032	0.129
BOD_5 (mg L^{-1})	0.255	0.012	**0.702**	0.056	-0.263	-0.098
Temperature (^{o}C)	-0.390	-0.246	**0.516**	0.428	-0.386	-0.083
pH	0.189	-0.293	-0.107	**0.785**	-0.278	-0.065
Chl a ($\mu g\ L^{-1}$)	-0.054	0.077	0.142	**0.773**	0.235	-0.168
DO (mg L^{-1})	0.005	-0.045	-0.158	0.118	**0.805**	0.115
TN (mg L^{-1})	0.080	0.269	0.371	0.161	**-0.545**	-0.382
TP (mg L^{-1})	0.093	0.269	0.232	-0.023	0.059	**0.790**
PO_4-P (mg L^{-1})	0.016	0.082	-0.194	-0.261	0.150	**0.700**
NO_3-N (mg L^{-1})	0.272	0.018	0.523	-0.113	0.502	-0.428
% Variance Explained	18.78	14.25	12.81	10.94	9.63	9.48

F denotes factors, N is number of cases and figures in bold are > 0.500, zoo - zooplankton

Table 2.4. Water quality in ponds, Period 3.

| | Gaba (n = 96) | | | | | | | | Walukuba (n = 180) | | | | | | | |
| | Pond 1 | | Pond 2 | | Pond 3 | | Pond 4 | | Pond 1 | | Pond 2 | | Pond 3 | | Pond 4 | |
	Mean	C.V.	Mean	C.V.	Mean	C.V.	Mean	C.V.	Mean	C.V.	Mean	C.V.	Mean	C.V.	Mean	C.V.
pH	7.4	13	8.3	9	8.2	8	8.1	8	8.5	8	8.7	9	8.8	10	8.8	8
Temperature (°C)	22.5	9	23.3	11	23.0	12	23.4	13	25.0	15	25.5	16	25.6	16	26.0	15
EC (μS cm^{-1})	285	17	434	23	419	14	488	23	1029	16	1251	27	1564	23	1611	12
DO	3.5	43	4.8	35	5.4	25	4.8	30	8.9	79	7.1	51	7.1	59	7.1	62
TSS	189	60	124	38	161	56	121	72	112	67	94	64	103	101	67	63
Turbidity (NTU)	410	131	174	77	249	90	192	85	97	107	84	87	86	109	57	96
Alkalinity (mg L^{-1})*	145	37	217	16	213	13	247	13	483	21	563	27	660	30	629	15
NO$_3$-N	0.140	236	0.132	347	0.155	249	0.027	244	0.043	558	0.014	574	0.011	611	0.235	366
NH$_4$-N	0.313	134	0.266	142	0.329	117	0.288	95	0.249	110	0.218	110	0.215	107	0.257	103
TN	1.55	132	1.85	125	1.75	92	1.26	143	2.14	167	1.55	125	1.95	104	2.52	185
PO$_4$-P	0.091	99	0.036	277	0.057	231	0.051	302	0.062	401	0.082	267	0.209	167	0.106	259
TP	0.472	78	0.308	109	0.337	90	0.261	52	0.664	58	0.555	62	0.816	80	0.691	115
Chl a (μg L^{-1})	37.5	243	42.8	137	61.8	166	34.0	201	94.9	125	60.2	131	69.4	143	62.1	222
Secchi depth (cm)	10.9	40	20.1	5	18.6	18	19.7	17	21.5	20	22.2	28	23.4	27	28.8	21
Pond volume (m^3)	235.7	10	224.0	11	152.8	10	255.4	10	275.2	21	159.8	28	172.5	44	301.0	12

Values represent means for twelve sampling dates in Gaba and seventeen in Walukuba and three sampling points (shallow, middle and deep) per pond, and Coefficients of variance (C.V.), n represents number of samples. * measured as CaCO$_3$

Table 2.5. Factor analysis results, pond water quality, Period 3.

Gaba (N = 96)	Factors			
Variable name	F1	F2	F3	F4
pH	**0.884**	−0.178	0.014	0.083
Secchi depth (cm)	**0.838**	0.139	0.035	0.376
Turbidity (NTU)	**−0.838**	−0.030	0.124	0.170
Alkalinity (mg L^{-1} as CaCO$_3$)	**0.803**	0.089	−0.078	0.399
DO (mg L^{-1})	**0.784**	−0.043	0.403	−0.075
TSS (mg L^{-1})	**−0.757**	0.103	0.404	−0.073
EC (µS cm^{-1})	**0.750**	0.246	−0.019	0.395
Temperature (°C)	0.100	**0.929**	0.014	0.084
NH$_4$-N (mg L^{-1})	−0.040	**0.720**	−0.001	−0.273
TN (mg L^{-1})	−0.034	**0.512**	0.319	0.173
Pond volume (m^3)	−0.220	0.131	**−0.775**	0.126
NO$_3$-N (mg L^{-1})	−0.452	0.133	**0.623**	0.109
Chl a (µg L^{-1})	−0.068	0.172	**0.594**	0.034
PO$_4$-P (mg L^{-1})	−0.042	0.312	−0.164	**−0.767**
TP (mg L^{-1})	−0.278	−0.235	0.184	**−0.676**
% Variance explained	32.88	13.29	12.34	11.27

Walukuba (N = 180)	Factors					
Variable name	F1	F2	F3	F4	F5	F6
EC (µS cm^{-1})	**0.913**	0.113	−0.111	0.116	0.033	0.020
Alkalinity (mg L^{-1} as CaCO$_3$)	**0.870**	−0.094	−0.003	−0.156	−0.010	−0.136
PO$_4$-P (mg L^{-1})	**0.500**	−0.172	−0.303	0.017	−0.272	0.001
NO$_3$-N (mg L^{-1})	−0.030	**0.832**	0.018	−0.019	0.149	−0.098
TN (mg L^{-1})	−0.060	**0.726**	0.037	0.144	−0.031	−0.108
Secchi depth (cm)	0.354	**0.627**	−0.325	−0.285	0.040	0.154
Pond volume (m^3)	−0.223	**0.570**	−0.236	0.201	0.090	0.448
TSS (mg L^{-1})	−0.141	−0.165	**0.846**	−0.011	0.088	−0.028
Turbidity (NTU)	−0.046	0.036	**0.829**	0.076	−0.292	−0.001
Temperature (°C)	0.311	−0.008	0.079	**0.809**	0.117	−0.063
DO (mg L^{-1})	−0.349	0.106	−0.012	**0.724**	−0.077	0.049
pH	0.074	0.024	−0.047	−0.142	**0.885**	0.025
NH$_4$-N (mg L^{-1})	−0.203	0.148	−0.133	0.330	**0.699**	−0.095
Chl a (µg L^{-1})	−0.300	−0.231	−0.168	0.002	−0.037	**0.655**
TP (mg L^{-1})	0.406	0.097	0.334	−0.112	−0.054	**0.640**
% Variance Explained	17.08	14.09	12.25	10.05	9.98	7.47

F denotes factors, N = number of cases. Loadings in bolded letters are > 0.500

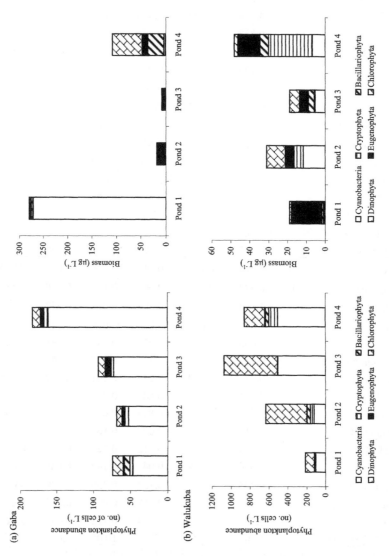

Figure 2.4. Phytoplankton abunduance and biomass in ponds, Period 3

Zooplankton species composition showed similarities between locations (Figure 2.5). The Arthropoda were from three Classes: Branchiopoda (Cladocera, 3 species); Maxillopoda (Copepoda, 1 order; copepodites, Cyclops, nauplii) and Ostractoda (1 order). Cyclopoida comprising mainly of naupli were dominant in both locations with Gaba having a higher proportion and 3 genera of Cladocera, *Ceriodaphnia*, *Diaphanosoma* and *Moina* were recorded. The Walukuba ponds had more Rotifera species with 12 genera against 8 for Gaba. Zooplankton biomass between July and September showed differences between the locations and within ponds with higher values in Walukuba. Pond 4 in Walukuba had consistently lower values.

Fish species and biomass in ponds, Periods 2 and 3
During the March-May rains in 2002 (Period 1), three species of *Oreochromis*, and *Protopterus aethiopicus* (Heckel, 1851) fry were found in pools of water at the Walukuba wetland-lake interface but ponds were not stocked. In Gaba, aided by a channel from the lake, three ponds were stocked with *Oreochromis niloticus* Linneaus 1758, *O. leucostictus* (Trewavas, 1933), *Haplochromines spp.* and *Aplocheilicthys pumulis* (Boulenger, 1906). No connectivity was achieved in Pond 1. Fish census in Gaba after ten months resulted in 282 kg ha^{-1} with fish weights from 2.0 to 187 g and fish lengths from 1 to 22.3 cm.

In the 2003 May-June floods (Period 3), the ponds in Gaba were stocked naturally with *O. niloticus, O. leucostictus, O. variabilis* (Boulenger, 1906) and *Haplochromine* species at both locations, *A. pumulis* in Walukuba only, and *Clarias gariepenus* (Burchell, 1822) and *Protopterus* in Gaba. In Walukuba, no connectivity was achieved for Pond 4, but 2 fish were found.

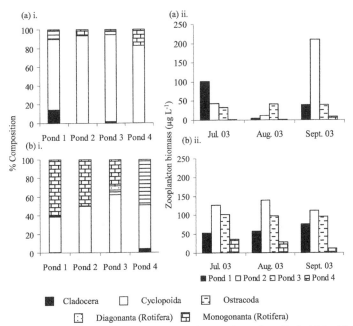

Figure 2.5 Zooplankton abundance and biomass by orders, Period 3 in (a) Gaba and (b) Walukuba

Table 2.8. Phytoplankton composition in ecotone (Period 2) and fish gut (initial stocking).

Group/Phyla Genera	Gaba				Walukuba			
	Lake	Wetland	Ponds	Fish	Lake	Wetland	Ponds	Fish
Cyanobacteria								
Anabaena	p	p	P	p	p	p	p	
Aphanocapsa	p		P		p		p	
Coelosphaerium	P		P		p	p	p	
Cosmarium	P	p	P		p	p	p	
Gomphosphaeria	P		P		p	p	p	
Merismopedia				p				
Microcystis	P	p		P	p	p	p	
Oscillatoria	P	p	P		p	p		
Planktolynbya				p				
Rhabdoderma	P						p	
Spirulina	P	p	P		p	p	p	
Chlorophyta								
Actinastrum			P		p			
Ankistrodesmus			p				p	
Chlamydella							p	
Chlamydomonas			p		p		p	
Chlorella	P	p	p		p	p		
Chroococcus	P				p			
Closterium	P		p		p	p	p	
Coelastrum				p				
Gonatozygon				p				
Pediastrum	P		p		p		p	
Scendesmus	P		p	p	p			
Sphaerellopsis	P		p		p		p	
Spirogyra	p	p			p	p	p	
Bacillariophyta								
Aulecoseira				p				
Navicula				p				p
Nitzschia				p				
Surirella				p				p
Tabellaria	p				p			
Euglenophyta								
Chrysophyta								
Euglena	p	p	p		p	p	p	
Monochrysis			p					
Phacus					p			p
Trachelomonas	p		p	p	p		p	
Dinophyta								
Peridinium				p				
Total Genera no.	**19**	**8**	**18**	**14**	**20**	**11**	**17**	**3**

P denotes present

O. leucostictus dominated both locations with 95% and 68% abundance in Gaba and Walukuba, respectively. *O. niloticus* was more abundant in Walukuba (20%). In both locations, fish sizes were variable with bigger fish (10-15 cm) in Walukuba and weights ranging from 2-40 g for the majority of fish. The initial fish biomass in Gaba was 92.3 kg ha^{-1} whilst in Walukuba it was 97.3 kg ha^{-1} (Table 2.9). Stocking densities ranged from 1.1 to 2.2 fish m^{-2} in both locations with exception of Pond 4's. The gut content of male and female *O. leucostictus* and *O. niloticus* (total length range 7.5-25.5 cm, 8-132 g) is shown in Table 2.8. They contained detrital material mainly comprised of sedimented algae; Bacillariophyta, Chlorophyta, Cyanobacteria, Dinophyta and Euglenophyta. *Chironomids* were also found in fish from both locations.

Table 2.9. Fish composition during initial stocking, period 3.

(a) Gaba

Variable	Pond 1	Pond 2	Pond 3	Pond 4
% number of fish stocking ponds				
Oreochromis niloticus	11.2	0.8	0	0
Oreochromis leucostictus	82.4	93.2	100	95.9
Protopterus aethiopicus	1.9	0	0	0
Haplochromine spp.	0.9	0	0	0
Aplocheilicthys pumulis	3.7	6	0	4.1
Total biomass at stocking (g)	3580	1714	1012	781
Population at stocking (n)	205	278	419	266
Fish Total length range (cm)	<5 - 20	<5 - 15	<5 - 15	< 5 - 15

(b) Walukuba

Variable	Pond 1	Pond 2	Pond 3	Pond 4
% number of fish stocking ponds				
Oreochromis niloticus	49.0	23.4	11.9	0
Oreochromis leucostictus	49	74	37.6	0
Protopterus aethiopicus	0.4	0	0	0
Haplochromine spp.	0.8	2.4	1.2	0
Aplocheilicthys pumulis	0.8	0	0	0
Total biomass at stocking (g)	2410	2455	2147	456
Population at stocking (n)	250	410	222	2
Fish Total length range (cm)	<5 - 15	<5 - 10	< 5 - 15	20 - 30

O. denotes *Oreochromis,* TL – total length.

Discussion

The ecotone and water quality

The productivity of natural papyrus swamps is variable and controlled by different factors such as climate, nutrient availability and the prevailing hydrological conditions. In this study the water quality of the two wetlands was similar to other natural swamps and did not show much variability within locations. pH and temperature were in ranges reported by Howard-Williams & Gaudet (1985) and Verhoeven (1986). DO concentrations were low mainly from the high oxygen demand exerted by decomposing organic matter which is typical of papyrus swamps (Carter, 1955; Gaudet, 1976). BOD values were relatively low comparable to the 2-10 mg L^{-1} reported for other natural swamps (Kadlec & Knight, 1996) which is attributed to the effective filtration of the thick papyrus root mats (Kansiime & Nalubega, 1999). Ammonia (NH_4-N) concentrations were also typical of papyrus swamps (Gaudet, 1975; Muthuri & Jones, 1997) though some high values were found. Total suspended solid (TSS) values were quite high (mean max. > 500 mg L^{-1}) particularly in Gaba which is attributed to differences in sediment texture; more sand in Gaba (60%) compared to Walukuba (44%).

Most of the variables examined in the wetland exhibited temporal variation hence the high C.V.s (greater than 100 %) (Figure 2.3) which arose from the season changes distinguished by the rainfall pattern (Figure 2.2). During the wetter months increments in concentrations of BOD, NH_4-N, turbidity and TSS are indications of nutrients being flushed from the wetland which is typical of papyrus swamps.

Five common processes described the variability of the water quality in the two wetlands namely buffering capacity, sedimentation/ re-suspension, oxygen, P release and organic matter decomposition. Using factor analysis the dynamics, regulation and processes of the wetland ecotones were explained. F1 showed a relation with alkalinity and EC (both positive) which indicates high total dissolved solids and more carbon dioxide respectively reflecting mineralization as well as a high buffering capacity. F2's in both wetlands were related to TSS which arose from settlement of particulate organic matter (organic/inorganic) and describes the sedimentation /re-suspension process. F3 Gaba and F5 Walukuba were both related to DO with the latter showing effects of distance and temperature which reflects higher productivity towards the open water. Furthermore, with high temperatures oxygen solubility is low. F4 Gaba and F3 Walukuba were related to PO_4-P pointing to P release. In Walukuba this was mainly a result of temperature effects while in Gaba impediment of PO_4-P occurred with increased pH as described by Delincé (1992) and Golterman (2004). Differences in causes of P release in the two locations are mainly explained by differences in soil texture. The silty sediment in Walukuba provides more sorption locations for P compared to the sandy sediment in Gaba (Golterman, 2004). F5 Gaba and F4 Walukuba were similarly related to BOD indicating the presence of organic matter thus describing the decomposition process. In Gaba F5 was also related to turbidity and NH_4-N reflecting sedimentation/re-suspension and subsequent release of NH_4-N whilst in Walukuba the factor was related to DO (negatively) and Cl⁻ (positively) which may suggest low sewage effluent pollution in this wetland.

The second period of the study was short covering only 8 weeks thus no meaningful temporal changes were observed hence the period focuses on spatial variation. Distinct differences in water quality between zones were found for several variables (Tables 2.4 & 2.5; Figure 2.6). Temperature, pH and DO were highest in the lake zone and lowest in the wetland in both locations while Chl *a* and zooplankton were highest in the lake zone in Walukuba and in the wetland zone in Gaba. In Walukuba these parameters point to active photosynthesis in the lake zone while in Gaba, due to the thick cover of water hyacinth (*Eichhornia crassipes*) in the inshore area, light limitation must have reduced algal productivity and hence zooplankton biomass. Furthermore, sequestration of nutrients by macrophytes limits phytoplankton productivity in wetlands as less nutrients are made available to algae. Low DO values in the wetland zone is a result of decomposition of macrophyte litter whilst the low variation in temperature in this zone (Figure 2.3) is attributed to shading by the thick macrophyte cover. Lehman *et al.* (1998) and Mnaya & Wolanski (2002) report similar differences between wetlands and lake waters. High nutrient concentrations (PO_4-P, TN, TP) and TSS in the wetland zone are attributed to high rates of litter fall, sedimentation and decomposition of organic matter (Verhoeven, 1986). Higher TN in the wetland zone is probably due to denitrification of the carbon rich anaerobic peat (Howard-Williams & Gaudet, 1985). High EC and alkalinity in the ponds is a result of the soil characteristics and to some extent climatic or seasonal changes as described by Boyd (1990) and reflect good buffering capacity in the ponds.

Water quality variability in the ecotone was explained by four main processes similar to those described for Period 1 namely: buffering capacity/zooplankton abundance as related to distance to the lake, sedimentation and re-suspension, organic matter decomposition and oxygen but in Walukuba two additional processes were found: photosynthesis and phosphorus release. For the first factor a strong correlation between distance and zooplankton abundance (Pearson correlation, $r^2 = -0.638$, $p<0.001$) was found for both locations. Since no correlation was found between zooplankton abundance and rainfall it is likely that flood levels had little effect on zooplankton abundance in the ecotone. Zooplankton is a critical food source and may determine how far fish can move within the wetland and explains the differences in fish types and size that stocked the ponds. In Gaba, carbon loss was mainly the result of organic matter decomposition (positive relation to NO_3 & TP) whilst in Walukuba rainfall had a bigger effect indicating flushing out of organic matter. Oxygen production in Gaba (F4) was regulated by pH while in Walukuba (F5) it was more related to TN suggesting mineralization of organic matter in anaerobic conditions. The remaining factors were stand alone processes with F5 in Gaba related to rainfall while F6 Walukuba was related to TP and PO_4-P indicating phosphorus release similar to findings in Period 1.

The distinction between the ecotone zones demonstrates that habitat specific characteristics play an important role in functionality of the ecotone (lake, wetland & ponds) as shown by Balirwa (1998) and in this study by the relationships between key processes. The relationship between buffering capacity and zooplankton abundance as related to distance (F1) and sedimentation/re-suspension (F2) shows a distinction between zones for F1 with ponds displaying the highest buffering capacity and lakes the least in both locations (Figure 2.6 (a)). The wider spread in the ponds in Walukuba is explained by the significantly higher alkalinity and EC values in these ponds and the variation between the ponds in comparison to Gaba while the stronger sedimentation/re-suspension effects on buffering capacity in the Gaba ponds are attributed to higher TSS in these ponds. The effects of oxygen on F1 are not so distinct but the wider spread in Walukuba signifies higher oxygen values in this site (Figure 2.6b). Effects of organic matter decomposition on buffering capacity are more distinct in Walukuba compared to Gaba suggesting more organic matter in this ecotone. With higher primary productivity in the Walukuba ponds more phytoplankton biomass is produced hence the increased organic matter within the pond system. For the wetland zone, a denser papyrus coverage is found in Walukuba compared to Gaba (Kaggwa et al., 2005) resulting in more litter decay hence more decomposition of organic matter within the wetland and subsequently the inshore lake area.

Plankton community densities in the lake and wetland zones gave insight into the initial plankton colonization of the ponds during flood events as well as indicated food availability for fish in these zones. Similar phytoplankton groups exist in both wetlands with relatively low species diversity and biomass attributed mainly to light limitation. Other researchers have found phytoplankton assemblages and productivity to change seasonally in response to variations in light environment, nutrient availability and differences in grazing pressures imposed by zooplankton and fish (Kling et al., 2001). Zooplankton play an important role in aquatic systems in providing food to many fish species as well as grazing on phytoplankton hence reducing its biomass. Their behaviour and population is regulated by seasonal and abiotic factors, predators, competition and abundance of phytoplankton. The number of individuals, their weights and development time are basic variables that are needed to estimate biomass and production (Bottrell et al., 1976). From the findings it is apparent that there is a horizontal shift of zooplankton towards the lake suggesting better food for the fish in this zone hence the attraction for fish to take refuge and spawn in these areas (Nalunkuma, 1998; Mbilinyi, 1998). Low zooplankton densities in the ponds in this study indicates their absence as a food source for the fish.

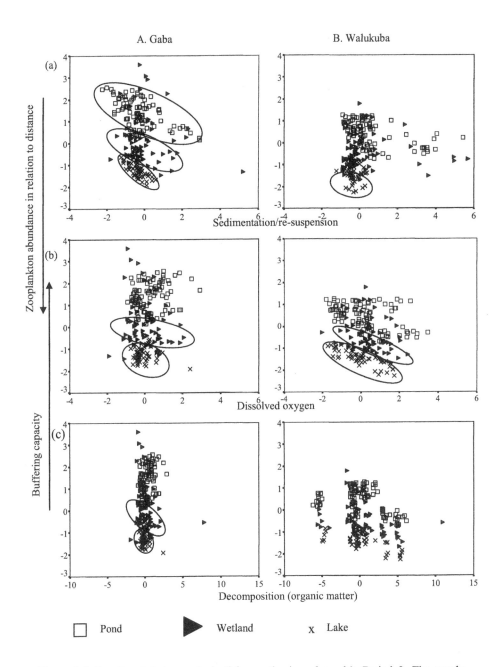

Figure 2.6. Results of factor analysis (lake, wetlands and ponds), Period 2. The graphs represent relationships between buffering capacity/zooplankton abundance as related to distance and (a) sedimentation/re-suspension (b) dissolved oxygen (DO) and (c) decomposition processes.

Despite similarities in the ecotone water quality in both locations, differences in pond water quality (Period 3) are more distinct. Temperature and pH show little variation between locations and between ponds similar to findings of Period 2. With exception of NO_3-N other variables were higher in Walukuba than in Gaba (Table 2.7). These differences are further confirmed by the results of the factor analysis. The first factor is related to pH, alkalinity and EC in both locations as found for Periods 1 and 2 but in Gaba the factor is also related to TSS and turbidity whilst in Walukuba the two processes are separated; F1 (buffering capacity) and F3 (sedimentation /re-suspension). F2 in Gaba is related to temperature and NH_4-N and to some extent TN pointing to the effect of temperature on ammonification whereas in Walukuba (F5) the release of NH_4-N depends more on pH. In Gaba pond volume has a strong effect on Chl *a* and NO_3-N values (F3) suggesting concentration of phytoplankton during low water levels whereas in Walukuba Chl *a* is more related to TP. F4 in Gaba is related to phosphorus release hence the high relation to TP and PO_4-P. Lastly F4 Walukuba is related to DO and temperature suggesting high primary productivity. The differences in the pond water quality and the processes describing its variability can best be explained by underlying factors such as soil characteristics (Table 2.1) and temperature.

Lotic and lentic fish community structure can be greatly influenced by physical disturbance such as floods and droughts (Freeman & Freeman, 1985). For populations and communities to persist in disturbed environments they seek refuge (Magoulick & Kobza, 2003); some within the wetland ecotone. Fish abundance in these areas may be related to habitat architecture, slope, food resources, predator-prey relationships, season and vegetation (Chick & McIvor, 1994; Balirwa, 1998). Mbilinyi (1998) found that access to the refugia by fish in the northern wetland ecotones of Lake Victoria depended on the characteristics of the habitat: sandy beaches had higher fish numbers and biomass compared to clay beaches. Low oxygen concentrations to which some species are very sensitive (Matthews, 1998) were found within the papyrus mats restricting habitation. Mbilinyi (1998) found *Oreochromis niloticus* mainly in *Phragmites* sites and *Oreochromis leucostictus* in the papyrus stands.

Water depth also influences the dynamics and structure of fish populations (Robinson & Tonn, 1989; Magoulick, 2000). The rising water levels in the wetlands during the flood periods result in fish migrating across the flood zone from the rich wetland-lake interface and from pools. Thus the importance of flood events (levels & duration) in natural stocking of Fingerponds. Water levels in 2003 in both Gaba and Walukuba were adequate to cause natural stocking of all ponds especially with fish fry and small fish (less than 10 cm). With dropping water levels in Lake Victoria of more than a metre in the last three years (GOU-MWLE, 2005) the likelihood of natural stocking of our trial Fingerponds system is bound to be affected seriously.

Fingerpond potential for fish culture

A conducive environment for the survival and culture of the fish, particularly tilapias, requires waters to have temperatures above 20°C (Rakocy & McGinty, 1989, Teichert-Coddington *et al.*, 1997), be well buffered by alkalinity with pH ranges 6.5-9.0 (Boyd, 1990; Wurts & Durborow, 1992) and not suffer from oxygen depletion (attain values above 2 mg L^{-1}). All these criteria are met in the study ponds. Ammonia, another parameter critical for fish survival was initially quite high but reduced with time. The reduction of NH_4-N in the water column may be attributed to phytoplankton development which utilizes the NH_4-N (Tucker *et al.*, 1984; Hargreaves, 1997). It is apparent that initial phytoplankton colonization of the ponds is crucial in driving primary production for transfer of energy up the food chain to fish in the ponds. However with high light attenuation due to high turbidity and TSS particularly in Gaba primary productivity in these ponds is likely to be limited. Furthermore, fish gut content indicated a restricted phytoplankton biomass in the ponds with fish found to graze mainly on detrital material. Initial zooplankton colonization of the ponds was rather poor and with low primary productivity, secondary productivity is likely to be limited. Additionally, zooplankton organisms in the ponds were mostly small (less than 0.1-3 mm) with mainly rotifers and development stages of copepods with adult Copepods, Cladocerans and Ostracods (0.5-1.0 mm) occasionally recorded. Porkorný *et al.* (2005) therefore concluded that only minute to medium sized zooplankton colonized the Fingerponds within a short time of flooding and that any bigger zooplankton were grazed upon by the fish.

Fluxes of nutrients and oxygen between mud and water play an important role in determining the water quality in a pond. Following pond construction, various processes begin to transform the bottom of a new pond into a pond soil (Boyd, 1990). The pond soil characteristics determined prior to the filling of the ponds (Table 2.1) showed clay content greater than 28 % which was comparable to other fish ponds in the region (Bowman & Seim, 1996; Boyd *et al.*, 1999), low organic matter content and low nutrients. A higher Ca content in Walukuba is attributed to the underlying rock and explains the higher alkalinity in Walukuba compared to Gaba. Pond soil characteristics change over time due to ageing, sedimentation and accumulation of N, P and organic matter (Boyd, *et al.*, 1994; Munsiri *et al.*, 1995). However in Walukuba the increment in clay content from 28.8 % in 2002 to 43.7 % in 2003 is highly unlikely within such a short time period. The increase is attributed to the clay base put in the ponds during reconstruction of the pond walls.

Pond embankments are subject to erosion (Steeby *et al.*, 2004) and can be a primary internal sediment load mechanism in ponds without turbid water supplies resulting in accumulation as ponds age (Munsiri *et al.,* 1995; Avnimelech *et al.*, 1999). This could explain the high turbidity and TSS concentration found in the Gaba ponds (Table 2.4). Organic matter from the sediment, settled algae and fish excreta may have negative impacts on the biological and chemical environment in the ponds. These get mineralized in the water column and at the sediment surface during the growing season (Boyd, 1995; 1997) which explains the absence of organic matter accumulation in the sediment in 2003 (Table 2.1).

Management options

Fish species that stock ponds determine to a great extent the success of their culture. In this study some variability in fish types that stocked the ponds were noted between the locations. In Walukuba, more diversity and bigger sizes (TL > 10 cm) stocked the ponds during flooding compared to the channel led Gaba ponds. *Oreochromis leucostictus* dominated both locations arising from its ability to withstand low oxygen concentrations (Mbilinyi, 1998). Overall, the fish survived in the ponds with no mortalities. With no manipulation, a fish yield of 282 kg ha^{-1} was attained in 300 days in Period 2 in Gaba. Natural stocking of ponds in 2003 resulted in an initial fish stock of 92.3 kg ha^{-1} and 97.3 kg ha^{-1} for Gaba and Walukuba respectively. In terms of food availability in the newly established ponds, fish gut content revealed fish to graze more on detritus material. This does not concur with previous findings where juvenile *Oreochromis* fry are reported to feed on zooplankton (Robotham, 1990; Porkorný *et al.*, 2005) and switch to phytoplankton with increase in size (Trewavas, 1983; Elhigzi *et al.*, 1995). This indicates the lack of adequate phytoplankton and zooplankton supply.

Key parameters that affect fish production are pH, available P and N, organic carbon and the C:N ratio. In this study, the low nutrient levels found in the ponds were adequate for fish survival but indicated a low potential for fish production as reported by Banerjea (1967). The potential of the finger pond system at these low nutrient levels can be estimated using theoretical computations based on the C:N:P ratio (Goldman, 1980) and the strong correlation between primary productivity and fish yield (Colman & Edwards, 1987). Taking the practical upper limit for net primary productivity of 4 g C m^{-2} d^{-1}, 8 kg N $ha^{-1}d^{-1}$ and 0.8 kg P $ha^{-1}d^{-1}$ are required to maintain this level. However, nutrient input in the pond required to achieve this productivity is expected to be less than the theoretical nutrient quota for the phytoplankton biomass. This is because the algal standing crop in fishponds is continuously or partially consumed by filter feeding fish and herbivorous organisms in the ponds. Therefore nutrients required only compensate algal growth and losses through grazing and sinking. Furthermore, only a small fraction of the nutrients ingested in the fish is assimilated, the largest amount is recycled back to the pond as waste (Lin *et al.*, 1997).

Based on these theoretical facts and the amounts of N and P found in the ponds, the predictable potential fish yields from this system with no further manipulation were calculated based on two assumptions (i) that N is available in the NO_3-N form and is the only limiting nutrient (ii) P is the limiting factor and PO_4-P the only available form. Using an ecological rule of the thumb that efficiency of energy transfer from one trophic level to the next is 10 % and that the carbon content in fish ranges from 10 to 16 % (Schroeder, 1987; Avnimelech & Lacher, 1979) the mean predicted fish yields ± SE using N as the limiting nutrient were 0.6 ± 0.4 t ha^{-1} yr^{-1}, n = 4 and for P as the limiting nutrient 1.2 ± 0.2 t ha yr^{-1}, n = 3.

Conclusions

Fringing papyrus wetlands on the northern shores of Lake Victoria provide a unique environment which allows for survival and movement of fish. Factors akin to fish survival namely; pH, ammonia, oxygen and temperature were conducive even in the drier seasons in the wetland ecotone and in the Fingerponds during the two grow-out periods. Key ecological processes that determined water quality variability in the ecotone and in the ponds were buffering capacity/zooplankton abundance and sedimentation/re-suspension. High turbidities and high suspended solids, if not controlled, may reduce fish production in these systems.

Natural stocking of Fingerponds is possible even when the ponds are not inundated and in drier periods channels can be used, provided that flood levels are adequate to fill the ponds. Fish types most likely to stock the ponds are *Oreochromis spp.* due to their tolerance of low oxygen levels but in low flood levels smaller sized fish (total lengths <5 cm) are more likely to stock the ponds. Low nutrient levels in the Fingerponds and low plankton biomass are not adequate for good fish production. With a high buffering capacity and high ranges of alkalinity, primary productivity can be greatly enhanced with the development of phytoplankton through nutrient input. This can be achieved through organic manure input which is more readily available and cheaper compared to chemical fertilizers.

Acknowledgements

This research was funded by the European Union, EU/INCODEV Fingerponds Project No. ICA4–CT–2001–10037. We are grateful to the 'Fingerpond Research group': UNESCO–IHE, Institute for Water Education, Delft, The Netherlands; ENKI, Czech Republic; Egerton University, Kenya; University of Dar es Salaam, Tanzania; Makerere University Institute of Environment and Natural Resources, Uganda; and Dr. Roland Bailey, UK for all the support and expertise. We are indebted to the management and staff of National Water and Sewerage Corporation, Uganda for providing technical support and facilities.

References

APHA, 1992. *Standard methods for examination of water and wastewater*, 18th Edition. American Public Health Association, Washington, DC.

APHA, 1995. *Standard methods for examination of water and wastewater*, 19th edition. American Public Health Association, Washington, DC.

Avnimelech, Y. and Lacher, M. 1979. A tentative nutrient balance in intensive fishponds. *Bamidgeh* 31.(1): 3-8.

Avnimelech, Y., Kochba, M. and Hargreaves, J.A. 1999. Sedimentation and re-suspension in earthen fish ponds. *Journal of the World Aquaculture Society* 30: 401-409.

Balirwa, J.S.1998. *Lake Victoria wetlands and the ecology of the Nile Tilapia Oreochromis niloticus Linné*. A.A. Balkema Publishers, Rotterdam, The Netherlands. 247 pp.

Banerjea, S.M. 1967. Water quality and soil conditions of fish ponds in some states of India in relation to fish production. *Indian. J. Fish*. 14: 113-144.

Bottrell, H.H., Duncan, A., Gliwicz M.Z., Grygierek, E., Herzig, A., Hillbright-Ilkowska A, Kurasawa H., Larsson, P. and Weglenska, T. 1976. A review of some of the problems in zooplankton production studies. *Norw. J Zool*. 24: 419-456.

Bowman, J.R. and Seim, W.K. 1996. Characterization of soils from potential PD/A CRSP sites in East Africa. Interim work plan, Africa study 4. 14th Annual Technical report. pp. 103-106.

Boyd, C.E. 1990. *Water quality in ponds for aquaculture*. Birmingham Publishing Co., Birmingham, Alabama. 482 pp.

Boyd, C.E. 1995. *Bottom soils, sediment and pond aquaculture*. Chapman and Hall. New York. 348 pp.

Boyd, C.E. 1997. Practical aspects of chemistry in pond aquaculture. *Progress Fish-Cult*. 59: 85-93.

Boyd, C.E., Queiroz, J. and Wood, C.W. 1999. Pond soil characteristics and dynamics of soil organic matter and nutrients. In: K. McElwee., D. Burke, Niles M. & Egna (eds.), 16th Annual technical report. *PD/A CRSP*, Oregon State University Corvallis. Oregon. pp 1-7.

Boyd, C.E., Tanner, M.E., Madkour, M. and Masuda, K. 1994. Chemical characteristics of bottom soils from freshwater and brackish water aquaculture ponds. *Journal of the World Aquaculture Society* 25: 517-534.

Carter, G.S. 1955. *The papyrus swamps of Uganda*. Heffer. Cambridge. 25 pp.

Chapman, D.C., Ehrhardt, E.A., Fairchild, J., Jacoson, R.B, Poulton , B.C., Sappington, L.C., Kelly, B.P and Mabee, W.R. 2004. Ecological dynamics of wetlands at Lisbon Bottom, Big Muddy National Fish and Wildlife Refuge, Missouri. Final Report to the US Fish & Wildlife service. Big Muddy National Fish and Wildlife Refuge, Columbia, MO. USGS. pp. 1-23.

Chick, J.V. and McIvor, C.C. 1994. Patterns in the abundance and composition of fishes among beds of different macrophytes. Viewing a littoral zone as a landscape. *Canadian Journal of Fisheries and Aquatic Sciences* 51: 2873- 2882.

Colman, J.A. and Edwards, P. 1987. Feeding pathways and environmental constraints in waste-fed aquaculture: balance and optimization. In: Moriarty, D.J.W. and Pullins, R.S.V. (eds.) *Detritus and microbial ecology in aquaculture*. ICLARM, Manila. 240 pp.

Delincé, G. 1992. *The ecology of the fish pond ecosystem with special reference to Africa*. Kluwer Academic Publishers, Dordrecht. 230 pp.

Denny, P. 1985 (ed.) *The ecology and Management of African Wetland vegetation*. Geobotany 6. Dr. W.Junk Publishers, Dordrecht, The Netherlands. 344 pp.

Denny, P. 1989. Wetlands. In: Strategic resources planning in Uganda. *UNEP* Report ix. 103 pp.

Denny, P. and Turyatunga, F. 1992. Uganda wetlands and their management. pp 77-84. In: Maltby, E., Dugan, P.J. and Lefeuvre, J.C. (eds.) 1992. *Conservation and development: The sustainable use of wetland resources.* Proceedings of the Third international wetland conference, Rennes, France, 19-23 September, 1988. ICUN, Gland, Switzerland, 219 pp.

Duncan, A. 1975. Production and biomass of three species of Daphnia co-existing in London reservoirs. *Verh. Internat. Verein. Limnol.* 19: 2858-2867.

Elhigzi, F.A.R., Haiderland, S.A. and Larsson, P. 1995. Interactions between Nile tilapia (O.n) and Cladocerans in ponds (Khartoum Sudan). *Hydrobiologia* 307: 263-272.

Fernando, C.H. (ed.) 2002. *A guide to tropical freshwater zooplankton. identification, ecology and impact on fisheries*. Backhuys Publishers, Leiden, The Netherlands. 291 pp.

Freeman, B.J. and Freeman, M.C. 1985. Production of fishes in a sub-tropical black water ecosystem: the Okefenokee swamp. *Limnology and Oceanography* 30: 686-692.

Gaudet, J.J. 1975. Mineral concentrations in papyrus in various African swamps. *Journal of Ecology* 63(2): 483-491.

Gaudet, J.J. 1976. Nutrient relationships in detritus of a tropical swamp. *Arch. Hydrobiol.*, 78: 213-239.

Goldman, J.C. 1980. Physical processes, nutrient availability and concept of relative growth rate in marine phytoplankton ecology. In: P.G. Falkowisk (ed.) *Primary productivity in the sea.* Plenum Press, New York. pp. 1-179.

Golterman, H.L. 2004. *The chemistry of phosphate and nitrogen compounds in sediments*. Kluwer Academic Publishers. 251 pp.

GOU MWLE, 2005 U*npublished data*. Meteorological database. Ministry of Water, Lands and Environment, Directorate of Water Development, Water Resources Management Department. Government of Uganda

Greenwood, P.H. 1966. *The fishes of Uganda* 2nd Edition. The Uganda society, Kampala. 131 pp.

Hargreaves, J.T.A. 1997. A simulation model of ammonia dynamics in commercial catfish ponds in the Southeastern United States. *Aquacultural Engineering* 16: 27-43.

Howard-Williams, C. and Gaudet, J.J. 1985. The structure and functioning of African swamps. In: Denny, P. (ed.). *The ecology and management of African wetland vegetation.* Dr. W. Junk Publishers, Dordrecht, The Netherlands. pp. 153-173.

Hyslop, E.J. 1980. Stomach contents analysis-a review of methods and their application. *Journal Fish Biology* 17: 411-429.

Jamu, D.M. and Ayinla, A. 2003. Potential for the development of aquaculture in Africa. *NAGA, World Fish Centre, Quarterly* 26 (3): 9-13.

Kadlec, R.H and Knight, R.L. 1996. *Treatment wetlands.* Lewis Publishers, Boca Raton. 893 pp.

Kaggwa, R.C., Kansiime, F., Denny, P. and van Dam, A.A. 2005. A preliminary assessment of the aquaculture potential of two wetlands located in the northern shores of Lake Victoria, Uganda. In: J. Vymazal (eds.) *Natural and constructed wetlands: Nutrients, metals and management.* Backhuys Publishers, Leiden, The Netherlands. pp. 350-368.

Kansiime, F. and Nalubega, M. 1999. *Wastewater treatment by a natural wetland: Nakivubo Swamp, Uganda.* Ph.D Thesis, A.A. Balkema Publishers, Rotterdam, The Netherlands. 300 pp.

Kling, H. J., Mugidde, R. and Hecky, R.E. 2001. Recent changes in the phytoplankton community of Lake Victoria in response to eutrophication. In: M. Munawar and Hecky R. E. (eds.) *The great lakes of the world (GLOW) Food web health and integrity.* Backhuys publishers Leiden. pp. 47-65.

Klute, A. 1986. Methods for soil analysis. *Part 1-Physical and mineralogical methods.* Am. Soc. Agronomy, Inc., Madison. 1188 pp.

Lehman, J.T., Mugidde, R. and Lehman, D.A. 1998. Lake Victoria plankton ecology: Mixing depth and climate-driven control of lake condition. pp. 99-116. In: Lehman, J. T. (ed.) *Environmental change and response in East African lakes.* Kluwer Academic Publishers, Dordrecht. 236 pp.

Lin, C.K., David, R., Teichert-Coddington, Green, B.W., and Veverica, K. L. 1997. Fertilization regimes. In: H.S. Egna and Boyd, C. E. (eds.) *Dynamics of pond aquaculture.* CRC Press LLC. Boca Raton, Florida. 73-104 pp.

Magoulick, D.D. 2000. Spatial and temporal variation in fish assemblages of drying steam pools: the role of abiotic and biotic factors. *Aquatic Ecology* 34: 29- 41.

Magoulick, D.D and Kobza, R.M. 2003. The role of refugia for fishes during drought: a review and synthesis. *Freshwater Biology* 48: 1186-1198.

Matthews, W.J. 1998. *Patterns in freshwater fish ecology.* Chapman and Hall, New York (second printing by Kluwer Academic Press). 756 pp.

Mbilinyi, H.G. 1998. Patterns of fish species composition and abundance in the Lake ecotone area of northern Lake Victoria, Jinja, Uganda, East Africa. M.Sc Thesis. DEW.056. IHE, Delft, The Netherlands. 50 pp.

Milstein, A. 1993. Factor and canonical analyses: basic concepts, data requirements and recommended procedures. pp. 24-31 In: M. Prein, G. Hulata & D. Pauly (eds.) *Multivariate methods in aquaculture research: case studies of tilapias in experimental and commercial systems.* ICLARM Stud. Rev. 20. Manila, Philippines, 221 pp.

Mnaya, B. and Wolsanki, E. 2002. Water circulation and fish larvae recruitment in papyrus wetlands, Rubondo Island, Lake Victoria. *Wetlands Ecology and Management.* 10: 133-143.

Munsiri, P., Boyd, C.E. and Hajek, B.F. 1995. Physical and chemical characteristics of bottom solid in ponds at Aurburn, Alabama, USA and a proposed system for describing pond soil horizons. *Journal of World Aquaculture Society.* 26: 346-377.

Muthuri, F.M. and Jones, M.B. 1997. Nutrient distributions in a papyrus swamp. Lake Naivasha, Kenya. *Aquatic Botany*. 56: 35-50.

Nalunkuma, G. 1998. Distribution patterns and ontogenetic flood shifts of juvenile and sub-adult *Tilapia zilli* and *Oreochromis niloticus* in the northern littoral zone of Lake Victoria, East Africa. M.Sc Thesis, IHE, Delft. The Netherlands. 49 pp.

Nauwerck, A. 1963. Die Beziehungen zwischen zooplankton und phytoplankton imsec. *Erhen.Symb. Bot. UpSal.* 17(5) 163 pp.

Novozamsky, I., Houba, V.J.G., Temminghoff, E. and Van der Lee, J.J. 1984. Determination of "Total N" and "Total P" in a single soil digest. *Neth. J. Agric. Sci.*32: 322-324.

Okalebo, J.R., Gathua, K.W. and Woomer, P.L. 1993. *Laboratory methods of plant and soil analysis: A Working Manual*. TSBF Programme. Nairobi, Kenya. 88 pp.

Pokornỳ, J. Přikryl, I., Faina, R., Kansiime, F., Kaggwa, R.C., Kipkemboi, J., Kitaka, N., Denny, P., Bailey, R., Lamtane, H.A. and Mgaya, Y.D. 2005. Will fish pond management principles from the temperate zone work in tropical fingerponds. In: J. Vymazal (ed.) *Natural and constructed wetlands: nutrients, metals and management*. Backhuys Publishers, Leiden, The Netherlands. pp. 382-399.

Prein, M. 2002. Integration of aquaculture into crop-animal systems in Asia. *Agriculture Systems* 71: 127-146.

Rakocy, J.E. and McGinty, 1989. Pond culture of tilapia. *Southern Regional Aquaculture Centre*. Publication No. 280. SRAC, Stoneville. M.S. 4 pp.

Robinson, C.L.K. and Tonn, W.M. 1989. Influence of environmental factors and piscivory in structuring fish assemblages of small Alberta Lakes. *Canadian Journal of Fisheries and Aquatic Sciences* 46: 81-89.

Robotham, P.W. J. 1990. Trophic niche overlap of fry and juveniles of *Oreochromis leucostictus* (Teleostei, Chichlidae) in the littoral zone of a tropical lake (L. Naivasha, Kenya). *Revue Hydrobiologie Tropicale* 23: 209-218.

Schroeder, G. L. 1987. Carbon and nitrogen budgets in manured ponds in Israel's coastal plain. *Aquaculture* 62: 259-279.

Steeby, J.A., Hargreaves, J.A., Tucker, C.S, and Kingsbury, S. 2004. Accumulation, organic carbon, dry matter concentration of sediment in commercial channel catfish ponds. *Aquacultural Engineering* 30: 115-126.

Teichert-Coddington, D.R., Popma, T.J. and Lovshin, L.L. 1997. Attributes of tropical pond-cultured fish. In: H.S. Egna and C.E. Boyd (eds.). 1997. *Dynamics of pond aquaculture*. CRC Press LLC. Boca Raton, Florida. pp. 183-198.

Trewavas, E.T. 1983. *Tilapiine fishes of the genera Sarotherodon, Oreochromis and Danakilia*. British Museum (Natural History) Publication, London UK. 583 pp.

Tucker, C.S., Lloyd, S.W. and Busch, R.L. 1984. Relationships between phytoplankton periodicity and the concentrations of total and un-ionized ammonia in channel catfish ponds. *Hydrobiologia* 111: 75-79.

Verhoeven, J.T.A. 1986. Nutrient dynamics in minerotrophic peat mires, *Aquatic Botany* 25: 117-137.

Wetzel, R.G. and Likens, G.E. 1991. *Limnological analyses*. 2nd Edition. Springer-Verlag. 391 pp.

Wurts, W.A. and Durborow, R.M. 1992. Interactions of pH, carbon dioxide, alkalinity and hardness in fish ponds. *SRAC Publication*. No. 464. 4 pp.

Chapter 3

Organic manuring in Fingerponds: effects on phytoplankton primary productivity and water quality and implications for management

Abstract

Eight 192 m^2 earthen ponds were constructed in two fringing wetlands on the northern shores of Lake Victoria and were stocked naturally with fish in the May to June rains of 2003. This study investigated the effect of manure inputs on natural food production and phytoplankton primary production and determined the subsequent changes in phytoplankton community structure and water quality between May 2003 and March 2005. Net phytoplankton primary productivity was significantly higher (P<0.05) in manured ponds compared to the unmanured ones and ranged from less than 0.01 to 28.3 g O$_2$ m^{-2} d^{-1}. Gross phytoplankton primary productivity was 1.1-2.4 fold the net primary productivity. The main limitations to phytoplankton productivity were high turbidity and decreasing pond water levels. Furthermore, low N:P ratios favoured the dominance of Cyanobacteria. During the culture period, dissolved oxygen did not drop below 2 mg L^{-1} at any one time and no oxygen depletion occurred. Despite the decreasing water levels over the culture period, the pond water quality did not deteriorate to levels unsuitable for fish survival but might have hampered fish growth.

Key words: Fingerponds, organic manure, phytoplankton, primary productivity, water quality.

Introduction

Fingerponds are earthen ponds dug at the edge of natural wetlands and are entirely dependent on natural flood events for the choice of fish densities and species (Denny *et al.*, 2006). Once the floods recede, the ponds are closed systems with water availability dependent on precipitation and seepage. Fish production in these ponds can then be enhanced by the use of natural, organic fertilizers.

Natural fertilizers have a long tradition of use in tropical aquaculture and represent an affordable source of nutrients for stimulating the algal and heterotrophic production cycles (Boyd, 1990). However, this dual effect on autotrophic and heterotrophic production is quite intricate. Fish feed largely on attached or planktonic algae, detrital/fungal flocs, or on zooplankton and macro-invertebrates (Delincé, 1992). Commonly cultured fish such as *Oreochromis* species switch from zooplankton (Robotham, 1990; Porkorný *et al.*, 2005) to phytoplankton (Trewavas, 1983) at a certain age and also utilize other natural food sources, hence the importance of both food chains.

Chicken manure has been used widely as an organic manure and has been found to be superior to other manures due to its high nutrient content (Green *et al.*, 1989; Delincé, 1992). However, semi-intensive aquaculture ponds that are over-manured develop dense phytoplankton populations that lead to photosynthetic light limitation (Smith & Piedrahita 1988; Knud-Hansen, 1997; Hargreaves, 1998).

Shallowness (thus ease of mixing) and high temperatures (enhancement of decomposition processes) are amongst the factors that facilitate nutrient release (Viner, 1969). However, following the application of fertilizers any unused material settles (Schroeder *et al.*, 1991), decomposes and can lead to anoxia (Hargreaves, 1998) and eventually fish kills. Nitrogen (N) and phosphorus (P) are often identified as limiting nutrients to algal biomass while silica is essential for diatom growth (Hecky & Kilham, 1988). Studies have shown that changes in nutrient loading affect phytoplankton productivity and community structure (Hecky & Kling, 1981; Carpenter *et al.*, 1985; Mugidde, 1993). Basic information on changes in water quality after the application of manure, and impacts on species composition, photosynthetic activity and phytoplankton community structure are vital for the success of fish culture. This study aimed at determining the relationships between manure applications, nutrient levels and phytoplankton production for optimization of fish production in order for recommendations to be made for the sound management of Fingerponds.

Materials and methods

Period, study area and experimental setup

This study was carried out from February 2003 to March 2005 in Gaba, Kampala (N 0° 14' 59.9", E 32° 38' 14.4") and Walukuba (near Jinja, N 0° 25' 58.1", E 33° 13' 59.8") on the northern shores of Lake Victoria, Uganda. In each site there were four Fingerponds (24 x 8 m, 1 m at the shallow end and 2 m at the deep end). The study covered two grow-out periods in two seasons. Season 1 (May 2003- January 2004) covered the period in which ponds were filled with flood water and stocked naturally with fish. In Season 2 (July 2004-March 2005) the ponds were only filled with rain and seepage water due to an unexpected drop in lake levels (ca. 1.5 m), further aggravated by poor rains. In Season 2, the ponds were functional from

October 2004 to January 2005 in Gaba and from November 2004 to March 2005 in Walukuba.

Chicken manure from local market stalls (Nitrogen, Phosphorus & Potassium values: 1.3-1.9%, 1.2-1.4%, 0.14-1.77%) was applied from August 2003 (Season 1) and October 2004 (Season 2) in Gaba and from October in both seasons in Walukuba. In Season 1, manure was applied to a bamboo cage in a pond corner fortnightly as follows: (a) Gaba, 521 kg ha^{-1} (low manure, LM) and 833 kg ha^{-1} (medium manure, MM) to ponds 1 and 2 respectively; (b) Walukuba, 521 kg ha^{-1} (LM), 833 kg ha^{-1} (MM) and 1563 kg ha^{-1} (high manure, HM) to ponds 2, 4 and 3 respectively. The other ponds acted as controls.

During Season 2, chicken manure and green biomass (1:5 ratio of chicken manure to green biomass - a mixture of grass and dried banana peals, composted for 3-5 days) were added to a bamboo cage in the corner of three ponds in each site at the same rate, namely 521 kg ha^{-1} (2 weeks)$^{-1}$ of chicken manure and one 10 L bucket of fermented green biomass for each pond on alternate weeks. Pond 4 (Gaba) and Pond 1 (Walukuba) acted as controls.

Sample collection (ponds)

In situ measurements were made monthly for pH, dissolved oxygen (DO), electrical conductivity (EC) and temperature using handheld meters (MODEL 340i, WTW, Weilheim, Germany). Sub-surface water samples were taken monthly from the shallow and middle of the ponds and at 10 cm vertical intervals from the deep end. Integrated samples collected from the deep end were used for phytoplankton determination. All measurements and sampling were done between 10.00 and 14.00 hours. Water and plankton samples were analyzed as described in subsequent sections.

Phytoplankton, primary productivity and photosynthetic parameters

Phytoplankton samples were preserved with 1% Lugol's solution. Sedimentation and enumerations were by inverted microscope following the modified Utermöl method (Nauwerck, 1963). Identification was to genus level (Mosille, 1994; John *et al.*, 2002). Numerical abundance scores were based on the DAFOR scale factor as described by Kent and Coker (1992). Chlorophyll *a* (Chl *a*) was extracted using 90% acetone: methanol (5:1 by volume) according to Pechar (1987) and biomass from biovolume measurements and subsequent calculations described by Wetzel and Likens (1991).

Mean photosynthetic active radiation (PAR) was derived from incident light radiation recorded every 5 minutes for diel measurements taken continuously using a light sensor (DataHog 2, Skye Instruments Ltd, Llandrindod Wells, UK). Underwater light was measured at 10 cm intervals using a light meter (SKP 200, Skye Instruments Ltd., Llandrindod Wells, UK). The euphotic depth was calculated from 1% surface light and vertical light attenuation coefficient from secchi depth (Parsons *et al.*, 1984).

Phytoplankton primary productivity was determined monthly using the modified Winkler's dark and light bottle method (Vollenweider *et al.*, 1974; Wetzel & Likens, 1991). Samples were taken from the top 30 cm. Incubations were carried out for 2 hours at 10-cm steps between 10.00 to 14.00 hours. The gross productivity attained for the photoperiod for the day was extrapolated with a conversion factor based on solar radiation integrated over the incubation period and the corresponding daily photoperiod. Net primary productivity (NPP), gross primary productivity (GPP) and respiration rates were computed from oxygen differences. Using a non-linear regression subroutine in SPSS 11.0 (SPSS Inc, Chicago, Illinois, USA) and Excel 2000 spreadsheet program, photosynthesis-light dependence curves were fitted to give the best R^2. The modified empirical equation offered by Platt et al., (1980) was used:-

$$P^B = P^B_{max}[1 - \exp(-\alpha^B I / P^B_{max})]$$ (Equation 3.1)

where P^B is the Photosynthetic intensity (mg C mg Chl $a^{-1} h^{-1}$) at the density of a light flux (μE m^{-2}); P^B_{max} is the maximum intensity of photosynthesis (mg C mg Chl $a^{-1} h^{-1}$); α^B is the initial slope of the light curve or the light utilization coefficient in mg C h^{-1} mg Chl $a^{-1} h^{-1}(\mu E$ $m^{-2})^{-1}$. Photosynthesis inhibition was assumed to be zero. The parameter I_k corresponds to the intersection of the extrapolated linear segment with the level of photosynthesis saturation, and was calculated as $I_k = P^B_{max} / \alpha^B$. Subscript B denotes photosynthetic rates normalized to Chl a. Calculations were based on monthly data.

Nutrient release experiment and subsequent effects on pond oxygen levels
The dispersion of ammonium nitrogen (NH_4-N) and soluble reactive phosphorus (SRP) were determined by their concentrations across a diagonal in the ponds 1, 6, 8, 15 and 22 m from the bamboo cage for day 1, 4, 7, 10, 13, 16 and 23 in Gaba and day 1, 3, 7, 14 and 21 in Walukuba at 12.00 hours following the first manure application in Season 1. Subsequent changes in dissolved oxygen (DO) concentrations were monitored in both locations at 06.00, 12.00, 14.00 and 18.00 hours on day 1-8, 11, 14, 16, 23, and 30 in Gaba and day 1, 3, 7, 14 and 21 in Walukuba. Chicken manure was applied as described above. Daily in situ measurements for pH, temperature, DO and EC were taken at 06.00, 12.00, 14.00 and 18.00 hours and for secchi depth at 12.00 hours. NH_4-N and SRP were determined as described below.

Analytical methods
Water samples were analyzed for ammonium nitrogen (NH_4-N) (APHA, 1992), total suspended solids (TSS), turbidity, nitrate nitrogen (NO_3-N), total nitrogen (TN), soluble reactive phosphorus (SRP), total phosphorus (TP), alkalinity and silica (SiO_2) according to standard methods (APHA, 1995) and as described in Chapter 2.

Statistical analysis
Statistical analysis was performed using SPSS 11.0 (SPSS Inc, Chicago, Illinois, USA). With no replicates, data were analyzed with multiple linear regression using an exploratory approach (Prein et al., 1993). Four groups of models were constructed, namely a) models relating pond inputs/conditions (explanatory variables) to individual water quality parameters (dependent variable); b) models relating pond inputs/conditions to primary production and phytoplankton; c) models relating water quality parameters to primary productivity and phytoplankton; and d) models relating time after application and distance from manuring cage to nutrient levels in the ponds to treatment. Significant regression

coefficients for the explanatory variables were interpreted as a significant effect of that variable on the dependent variable.

Data used were for both locations and for both seasons. For the different sets of models the numbers of cases used were: for a) N=1303 cases for DO measurements from the sub-surface across a diagonal, N=264 for SRP and NH_4-N. For models b, c and d, 96 cases (44 cases for Gaba, 52 cases Walukuba) comprised of monthly measurements in Seasons 1 and 2 (8 and 3 months in Gaba; 9 and 4 months in Walukuba) made up each data set.

A complete correlation matrix was accomplished for all data sets to determine if predictor variables were correlated which would pose a danger of multi collinearity (Hopkins *et al.*, 1988). Prior to estimating the models, normality of the data was determined. To stabilize the variance, transformations were made for TSS, turbidity, NO_3-N, SRP, N:P ratio, DO, alkalinity, GPP, NPP, Chl *a*, algal biomass and algal density (natural logarithm), TN (inverse) and TP (negative exponential). Dummy variables were created for season (Season 1 = 1, Season 2 = 0), location (Gaba = 1, Walukuba = 0) and green biomass (1 = green biomass added to pond, 0 = green biomass not added to pond). Other variables introduced included manure load and pond water level. Standardized partial regression coefficients were computed for all models as described in Yamane (1973). For every model, the Durbin-Watson statistic (DW) was calculated to test for positive or negative serial correlation. Significance levels for independent variables were 5 % (*) 1 % (**) and 0.1 % (***).

Results

Water quality, seasons 1 and 2
Physical parameters
In the first season, Season 1, all ponds at their mid-point filled to depths from 0.8-1.1 m but in the second season, Season 2, due to drought they rarely exceeded 0.5 m and sometimes were considerably lower (Figure 3.1). Variations between locations and differences in the pond environments in the two seasons are given in Tables 3.1a and b. Water temperatures varied between 22.8 and 29.0 °C. pH values ranged from neutral to alkaline (up to nearly pH 10), the ponds with manure generally being more alkaline. Conductivity values ranged from 309-1909 μS cm^{-1} with the higher values in Season 2. The ponds in Walukuba overall had higher conductivities than those in Gaba which corresponded with approximately 2-fold higher alkalinity values (up to 1228 mg $CaCO_3$ L^{-1}). Total suspended solids (TSS) and turbidities were high, rising to 950 mg L^{-1} and greater than 500 NTU resulting in low secchi disc readings of less than 30 cm.

Oxygen levels
Over the two seasons, DO concentrations never dropped below 4.0 mg L^{-1} for measurements taken between 10.00 and 14.00 hours but were distinctively higher in Walukuba in Season 1 (Figure 3.2) with a value of 12.8 mg L^{-1} in the highly manured pond. Diel measurements for DO for the Gaba ponds, taken prior to manure application in June 2003 and later after manuring in January 2004, followed typical cycles with no major differences between the two sampling dates (Figure 3.3). Minimum DO concentrations for these dates were always less than 2 mg L^{-1}, both at the surface and bottom.

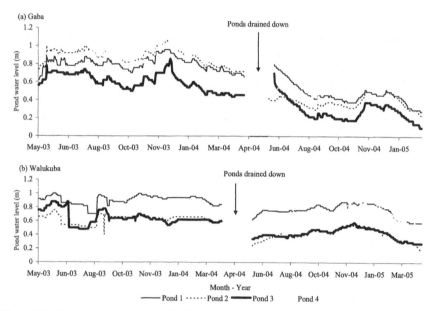

Figure 3.1. Pond water levels (m) in Gaba (May 2003 to February 2005) and Walukuba May 2003 to April 2005). Values plotted are daily readings taken in the pond at 12.00 hours.

Nutrients and chlorophyll *a*

After the application of manure, nutrient concentrations were variable. NO_3-N concentrations were less than 0.05 mg L^{-1} and showed slight increment with manure input whereas ammonium nitrogen (NH_4-N) concentrations were initially high in both locations but decreased with time. In the second season, S2, NH_4-N concentrations rose to a high of 9 mg L^{-1} in Gaba at pH 8.7 but later dropped. Highest NH_4-N levels in June occurred at 22.00 hours in the unmanured ponds while in January they occurred at 08.00 hours in the unmanured ponds and at 20.00 hours in the manured ponds (Figure 3.3). TN and TP increased following manure application (Figure 3.4) while silica did not vary and ranged between 5 and 52 mg L^{-1} with values greater than 15 mg L^{-1} found in Walukuba. Chlorophyll *a* content increased over time proportionally to manure input levels and ranged from 10 to 730 µg L^{-1} with the higher concentrations recorded in Walukuba.

Table 3.1a. Water quality analysis, Gaba seasons 1 and 2

Parameter	Season 1						Season 2			
	Pond 3 & 4, NM n = 20		Pond 1, LM n = 6		Pond 2, MM n = 6		Pond 4, NM n = 12		Pond 1, 2 & 3, LM n = 9	
	Mean	C.V.	Mean	C.V.	Mean	C.V.	Mean	C.V.	Mean	C.V.
pH	8.3	4	8.1	3	8.4	3	8.7	3	8.6	2
EC (μS cm^{-1})	435	13	351	5	459	4	755	9	577	9
Alkalinity (as $CaCO_3$)	219	2	177	0	220	0	340	9	242	0
DO	5.0	34	4.8	22	5.1	0	5.5	41	6.4	26
Temperature (°C)	24.4	8	24.1	5	24.7	5	24.2	7	25.8	6
Chl a (μg L^{-1})	31.9	235	53.3	72	61.2	39	119.5	47	140.7	21
NH_4-N	0.314	70	0.248	24	0.236	37	3.736	20	1.281	28
NO_3-N	0.101	39	0.036	27	0.161	1	0.027	74	0.008	221
Secchi depth (cm)	16.7	15	17.7	8	19.2	0	11.3	13	15.9	1
TSS	142	36	93	21	85	42	344	15	161	29
Turbidity (NTU)	177	27	136	21	90	38	413	23	157	25
SiO_2	18.6	0	22.2	0	15.3	11	14.4	12	12.0	0
SRP	0.039	69	0.061	27	0.083	10	0.054	47	0.069	47
TN	3.1	125	3.2	41	3.9	63	7.6	31	13.8	24
TP	0.330	214	0.530	19	0.630	80	0.872	52	1.588	17

All units in mg L^{-1} unless otherwise stated. Values represent means and correlation of variance (C.V.s) for water samples taken monthly from the sub-surface shallow and middle pond ends, and vertical profiles at deep ends. NM denotes no manure, LM – low manure, MM – medium manure and HM – high manure, n = number of samples.

Table 3.1b. Water quality analysis, Walukuba seasons 1 and 2

Parameter	Season 1								Season 2			
	Pond 1, NM		Pond 2, LM		Pond 4, MM		Pond 3, HM		Pond 1, NM		Pond 2, 3 & 4; LM	
	n = 24		n = 4		n = 4		n = 4		n = 5		n = 15	
	Mean	C.V.	Mean	C.V.	Mean	C.V.	Mean	C.V.	Mean	C.V.	Mean	C.V.
pH	8.8	3	9.0	0	9.1	2	9.2	1	8.9	1	9.2	2
EC ($\mu S\ cm^{-1}$)	1251	2	1256	81	1633	1	1570	0	1030	2	1572	2
Alkalinity (as $CaCO_3$)	557	0	613	5	678	4	670	0	527	18	816	0
DO	7.4	33	7.6	12	6.7	27	9.7	20	5.4	33	6.9	49
Temperature (°C)	26.6	4	28	2	28.2	2	28.1	2	25.7	0	26.2	6
Chl a ($\mu g\ L^{-1}$)	114.5	137	92.0	53	59.6	28	171.0	44	112.2	117	418.0	29
NH_4-N	0.237	47	0.268	18	0.228	16	0.385	18	0.837	11	1.047	13
NO_3-N	0.040	150	0.000		0.000		0.000		0.007	337	0.015	506
Secchi depth (cm)	21.8	3	18.6	5	26.2	4	18.6	0	18.9	8	13.5	11
TSS	122	68	172	48	59	118	196	18	103	23	136	23
Turbidity (NTU)	75	51	80	37	34	163	74	19	74	25	79	26
SiO_2	34.1	0	20.2	0	19.5	0	20.7	11	23.1	17	19.8	0
SRP	0.047	37	0.186	2	0.228	13	0.456	20	0.031	349	1.729	19
TN	3.5	134	4.1	40	4.1	19	6.7	67	29.1	12	36.5	30
TP	0.698	45	1.020	32	1.028	19	2.440	17	0.782	37	6.046	16

All units in mg L^{-1} unless otherwise stated. Values represent means and correlation of variance (C.V.s) for water samples taken monthly from the sub-surface shallow and middle pond ends, and vertical profiles at deep ends. NM denotes no manure, LM – low manure, MM – medium manure and HM – high manure, n = number of samples

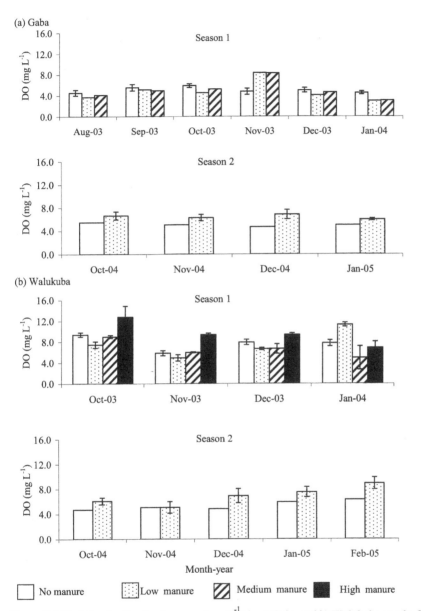

Figure 3.2 Variation in dissolved oxygen (mg L^{-1}) in a) Gaba and b) Walukuba ponds, Season 1 (S1) and Season 2 (S2) (treatments applied). Charts represent monthly means for measurements while error bars represent standard error of the mean for ponds for samples taken from the shallow, middle and deep end of the ponds.

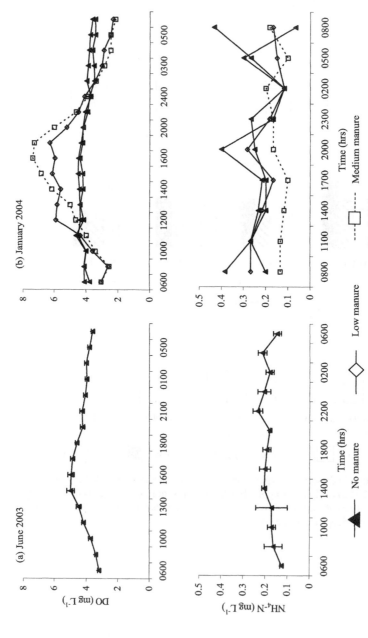

Figure 3.3. Diel cycles for dissolved oxygen (DO, mg L⁻¹) and ammonium nitrogen (NH₄-N, mg L⁻¹) in the Gaba ponds in a) June 2003 (no manure) and b) January 2004 (no manure, low manure and medium manure). Values plotted are measurements taken at specific times while bars represent standard errors of the mean and indicate variation between ponds with the same treatment.

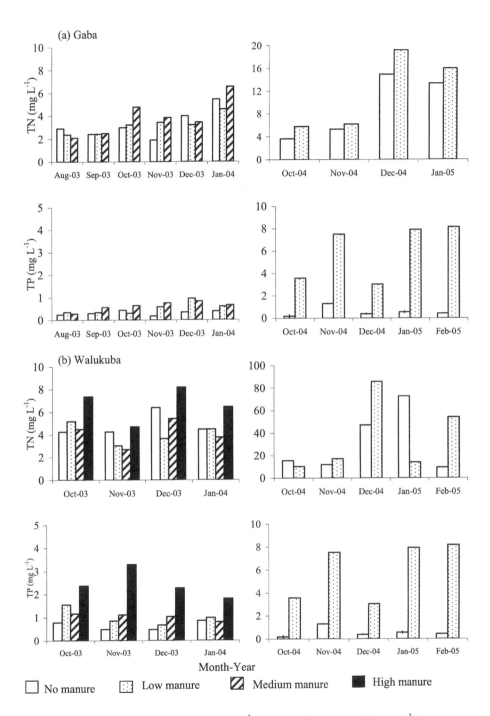

Figure 3.4. Variation in total nitrogen (TN, mg L^{-1}) and total phosphorus (TP, mg L^{-1}) in ponds, seasons 1 and 2.

Effects of pond inputs/conditions on water quality

Regression results for effects of pond inputs/conditions on water quality are shown in Table 3.2. There were three independent variables: the dummy variables location and season, and manure load. The 14 dependent variables were EC, temperature, SiO_2, DO, TSS, turbidity, alkalinity, NO_3-N, NH_4-N, TN, SRP, TP and pH (all from field measurements) and N: P ratio (computed). The final data set consisted of 96 cases with 17 variables. All models were significant (P<0.001), except the model for TSS. The models for EC and alkalinity explained the highest amount of variance (> 80 %) while the other models generally explained 20-60%. The models for SiO_2, NO_3-N and TSS explained less than 20% of the variation in the dependent variable.

In general EC, temperature, SiO_2, DO, turbidity, alkalinity, ammonia, N:P, TP and pH were significantly different (P<0.05) in the two locations while no significant differences were noted for TSS, NO_3-N, TN and SRP. Higher values of EC, temperature, SiO_2, DO, alkalinity, SRP, TP and pH were attained in Walukuba (negative beta weights) in comparison to Gaba. Turbidity, TSS, NO_3-N, NH_4-N, TN and N:P were higher in Gaba. EC, alkalinity, NH_4-N, TN and TP were higher in Season 2 than Season 1 while NO_3-N was higher in Season 1. The other variables were unaffected by season.

Manure input had a significant and positive effect on EC, temperature and nutrients (SRP, TN, N:P, TP) and a lowering effect on turbidity levels and pH.

Phytoplankton primary productivity and biomass

Phytoplankton net primary productivity (NPP) and the algal biomass measured as Chlorophyll a was enhanced in both locations in the first season following manure input (Figure 3.5). In Gaba, the medium manured pond, MM, attained the highest NPP values compared to the other ponds and increased steadily over the season while in Walukuba neither distinct differences nor trends were noticed between ponds. Mean NPP in Gaba ranged from less than 0.1 to 4.9 g O_2 m^{-2} d^{-1} and from less than 0.1 to 28.2 g O_2 m^{-2} d^{-1} in manured and unmanured ponds, respectively while in Walukuba it was 0.3-16.3 and 0.60-16.0 g O_2 m^{-2} d^{-1}, respectively. The highest productivity was attained in the upper 30 cm layer of the water column. During Season 2, a general decrease in primary productivity levels was observed and surprisingly in Walukuba, the un-manured pond was as productive as the manured ones. Gross to net primary productivity ratios (GPP:NPP) ranged between 1.1 and 2.4.

Chl a concentrations increased with manure input from less than 10 μg L^{-1} to over 100 μg L^{-1} in Gaba while in Walukuba values rose to 250 μg L^{-1}. In Season 2, higher Chl a concentrations were obtained with lowering pond water levels; 623 and 700 μg L^{-1} in Walukuba and Gaba, respectively. A linear relationship was found between Chl a and TP with a higher correlation found in Gaba (R^2; 0.40 to 0.43) in comparison to Walukuba (R^2; 0.06 to 0.08) and differences found between seasons. In Gaba, higher values of TP were obtained in Season 2 which resulted in a proportionate increment in Chl a concentrations.

The relationship between Chl a and secchi depth showed some variability for both locations but between TSS and Chl a more variability was demonstrated in Gaba (Figure 3.6). The low pond transparencies were reflected in the light variables: euphotic depth ranged from 27-107 cm in Gaba and 32-109 cm in Walukuba with the deeper ponds 4 (NM, Gaba; MM, Walukuba) attaining higher euphotic depths. Differences between seasons were noted with lower values in Season 2 for both locations. Vertical light attenuation coefficients for both seasons ranged from 8.4-16.2 m^{-1} in Gaba and 6.5-11.2 m^{-1} in Walukuba.

The effect of inputs and conditions on phytoplankton and primary productivity are shown by the regression models in Table 3.3. All models were significant ($P<0.001$) but models for NPP and GPP accounted for less than 10 % of the overall variance whilst the other models accounted for 11-37%. Lower values of secchi depth, algal densities and Chl a were attained in Gaba. Higher secchi depths occurred in Season 1 while algal densities and Chl a were lower in Season 1 than in Season 2 (negative beta weights). NPP and GPP were both significantly and positively affected by manure load.

Water quality effects on primary productivity and phytoplankton are shown by the regression models in Table 3.4. All models were significant ($P<0.001$) and explained 20- 48% of the overall variance with secchi depth and Chl a explaining more than 40%. Secchi depth, NPP and GPP were significantly and negatively affected by TSS which was ranked their second most important explanatory variable. Phytoplankton density, Chl a, NPP and GPP were related positively to TP which was ranked the most important explanatory variable for these parameters. Secchi depth was also positively and significantly affected by alkalinity whilst NH_4-N and SiO_2 had a negative effect.

Phytoplankton photosynthetic parameters
Results for this section are presented in Table 3.5 and are based on data collected over the two seasons. Maximum intensity of photosynthesis (P^B max) was generally higher in Walukuba compared to Gaba in both seasons. Light utilization coefficients (α) in Walukuba did not differ much between manured and unmanured ponds. In Walukuba the low manured, LM, and medium manured, MM, ponds in Season 1 and LM pond 3 did not reach saturation. In Gaba α showed no distinct pattern between manured and unmanured ponds in both seasons whilst light saturation was higher in Walukuba..

Changes in phytoplankton species composition and biomass, seasons 1 and 2
Six phytoplankton phyla rich in genera diversity were identified in all ponds: Bacillariophyta, Chlorophyta, Cryptophyta, Cyanobacteria, Dinophyta and Euglenophyta (Table 3.6 a & b). The highest genera diversity was present in the Cyanobacteria group (16 genera) followed by Chlorophyta (15 genera). In Gaba, a lower genera diversity was found in Season 2 whereas in Walukuba the diversity increased. Following flooding and stocking of ponds at the onset of the first season, Season 1, a similar phytoplankton community was found in all ponds in Walukuba but not in Gaba where during the months of June and July the phytoplankton community was more pond-specific (Figure 3.7). Differences in phytoplankton biomass between ponds were noted in both locations with higher biomasses attained in the manured ponds for both seasons. Phytoplankton biomass ranges in manured ponds were 1,928-71,852 mg m^{-3} in Gaba and 1,356-311,873 mg m^{-3} in Walukuba while in unmanured ponds they were 204-242,749 mg m^{-3} in Gaba and 137-22,008 mg m^{-3} in Walukuba. Phytoplankton densities too showed a cyclic pattern over time (Figure 3.8).

Table 3.2. Multiple regression models for pond inputs/ conditions against water quality variables.

Model	Dependant variable		DV location	DV Season	Manure load	Adj. R^2	F	Constant
1	EC (μS cm^{-1})	B	-877.960	-110.081	4.196	0.826	151.629 ***	1408.300
		beta	-0.870 ***	-0.110 *	0.114 *			
2	Temperature (°C)	B	-1.780	-0.208	0.024	0.333	16.830 ***	26.441
		beta	-0.519 ***	-0.055	0.191 *			
3	SiO$_2$ (mg L^{-1})	B	-7.323	2.367	-0.148	0.106	4.771 **	23.389
		beta	-0.342 **	0.101	-0.190			
4	LN DO	B	-0.328	-0.020	-0.001	0.265	12.389 ***	1.976
		beta	-0.543 ***	-0.031	-0.049			
5	LN TSS	B	0.195	-0.104	-0.000	0.008	1.271	4.727
		beta	0.180	-0.088	-0.011			
6	LN Turbidity	B	0.650	-0.148	-0.010	0.285	13.639 ***	4.445
		beta	0.469 ***	-0.098	-0.205 *			
7	LN Alkalinity	B	-1.016	-0.115	0.002	0.871	214.469 ***	6.469
		beta	-0.913 ***	-0.094 *	0.054			
8	LN NO$_3$-N	B	0.120	0.900	-0.015	0.076	3.615 *	-4.714
		beta	0.040	0.276 **	-0.136			
9	Inverse NH$_4$-N	B	-0.564	2.853	-0.016	0.518	35.087 ***	0.574
		beta	-0.154 *	0.709 ***	-0.115			
10	Inverse TN (mg^{-1} L^{-1})	B	-0.063	0.328	-0.005	0.392	21.377 ***	0.142
		beta	-0.122	0.576 ***	-0.256 **			
11	LN SRP	B	-0.594	-1.255	0.050	0.461	28.031 ***	-1.839
		beta	-0.201	-0.388 ***	0.466 ***			
12	LN N:P	B	0.659	-0.711	-0.017	0.287	13.737 ***	2.149
		beta	0.341 ***	-0.337 ***	-0.248 **			
13	e^{-TP}	B	0.195	0.276	-0.007	0.595	47.567 ***	0.223
		beta	0.377 **	0.489 ***	-0.362 ***			
14	pH	B	-0.578	0.004	-0.212	0.500	32.719 ***	9.043
		beta	-0.637 ***	0.126	-0.213 **			

DV denotes dummy variable. Independent variables are: Dummy variable location, dummy variable season and manure load (kg ha^{-1} (2weeks)$^{-1}$). Significant levels for partial regression coefficients are marked * 5% . ** at 1% and *** at 0.1 %. N = number of cases = 96.

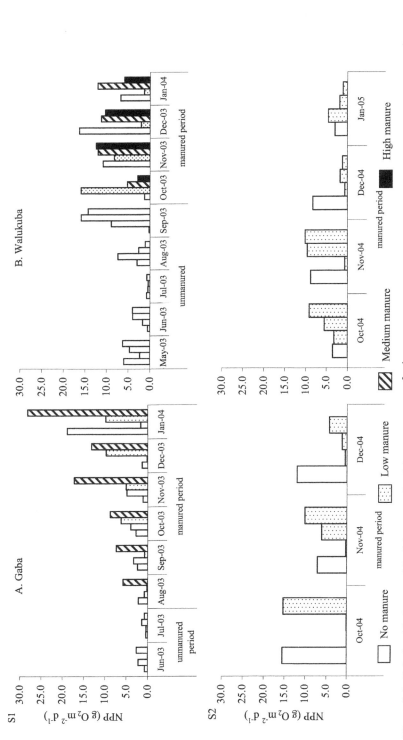

Figure 3.5. Relationship between Net Primary productivity (NPP, g O$_2$ m^{-2} d^{-1}) and time for A. Gaba and B. Walukuba ponds, seasons 1 and 2 (S1 and S2). Values represent measurement taken monthly.

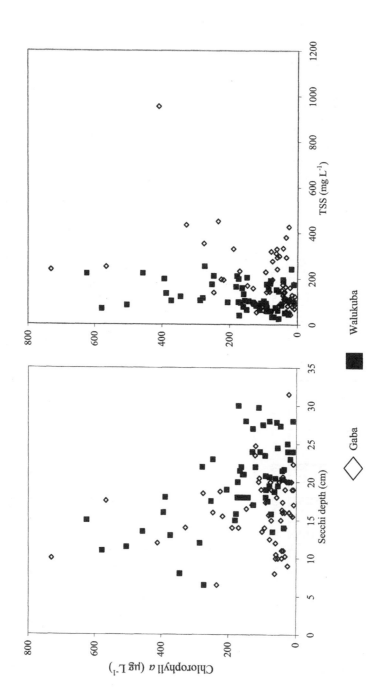

Figure 3.6. Relationship between Chlorophyll *a* (Chl *a* µg L⁻¹) and secchi depth (cm) and total suspended solids (TSS, mg L⁻¹), seasons 1 and 2.

Table 3.3. Multiple Regression models for pond inputs/conditions against phytoplankton primary productivity and density.

Dependent variables		DV Location		DV Season		Manure load	Adj. R²	F		Constant
Secchi depth (cm)	B	-3.622		5.647		-0.017	0.362	25.988	***	6.831
	beta	-0.363	***	0.517	***	0.048				
LN Algal density	B	-0.869		-0.797		-0.005	0.109	4.858	**	19.497
	beta	-0.283	**	-0.237	*	-0.043				
LN Chl a	B	-0.748		-0.996		0.006	0.372	19.739	***	5.254
	beta	-0.378	***	-0.459	***	0.081				
LN NPP	B	-0.370		0.257		0.029	0.078	3.666	*	0.798
	beta	-0.126		0.064		0.273	*			
LN GPP	B	-0.244		0.028		0.026	0.096	4.346	**	1.384
	beta	-0.108		0.011		0.313	*			

Adj. denotes Adjusted, Chl a-chlorophyll a, R-regression coefficient, DV-dummy variable, F-F statistic, NPP- net primary productivity, GPP- gross primary productivity. Independent variables are: Dummy variable location, dummy variable green biomass and manure load (kg ha⁻¹ (2weeks⁻¹)). Significant levels for partial regression coefficients are marked * at 5 %, ** at 1% and *** at 0.1 %, N = number of cases = 96.

Table 3.4. Multiple regression models for water quality variables against phytoplankton primary productivity and density.

Dependent variables		LN TSS	e^{-TP}	LN N: P	INV NH$_4$-N	SiO$_2$	LN SRP	LN Alk	Adj. R^2	F	Constant
Secchi depth (cm)	B	-2.743	4.746	-0.508	0.727	-0.075	-0.517	3.763			5.771
	beta	-0.297 ***	0.246	-0.112	0.268 **	-0.161 *	-0.153	0.420 ***	0.475	13.264 ***	
LN Algal density	B	-0.551	-4.18	0.195	0.190	0.008	-0.123	-0.332			23.527
	beta	-0.194	-0.703 ***	0.122	0.227	0.056	-0.118	-0.120	0.197	4.324 ***	
LN Chl a	B	0.113	-2.246	0.131	-0.036	-0.003	0.024	0.200			3.551
	beta	0.062	-0.585 ***	0.127	-0.067	-0.033	0.036	0.113	0.402	10.105 ***	
LN NPP	B	-1.065	-2.75	0.338	0.159	-0.01	-0.097	-0.046			6.464
	beta	-0.391 ***	-0.482 **	0.221	0.198	-0.092	-0.097	-0.017	0.203	4.460 ***	
LN GPP	B	-0.955	-2.458	0.109	0.048	-0.004	-0.088	-0.324			8.569
	beta	-0.456 ***	-0.561 **	0.093	0.077	-0.041	-0.116	-0.159	0.257	5.684 ***	

Adj. denotes Adjusted, INV-inverse, LN-natural logarithm, TSS-total suspended solids, TP-total phosphorus, NH$_4$-N-ammonium nitrogen, SiO$_2$-silica, SRP-soluble reactive phosphorus, Alk-alkalinity, R-regression coefficient, F-F statistic, Chl a-chlorophyll a, NPP-Net primary productivity, GPP-gross primary producti vity. Significant levels for partial regression coefficients are marked * at 5 %, ** at 1% and *** at 0.1 %, N = number of cases = 96.

Table 3.5. Summary of monthly phytoplankton photosynthetic parameters computed for seasons 1 and 2.

(a) Gaba

Manure level (pond no)	R^2	P^B_{max}	SE	α^B	SE	I_k
Season 1						
NM (pond 3)	0.20	14.0	3.6	0.40	1287.0	35
NM (pond 4)	0.61	122.2	14.1	1.59	2556.0	77
LM (pond 1)	0.37	46.5	4.0	1.03	17773.0	45
MM (pond 2)	0.97	72.3	17.3	0.05	43.7	1341
Season 2						
NM (pond 4)	*	25.3	36.6	1.29	26104.0	20
LM (pond 1)	0.62	25.5	18.0	0.16	367.8	163
LM (pond 2)	0.95	24.3	2.1	0.10	63.4	246
LM (pond 3)	0.90	41.3	12.2	0.06	74.6	733

(b) Walukuba

Manure level (pond no)	R^2	P^B_{max}	SE	α^B	SE	I_k
Season 1						
NM (pond 1)	0.98	39.0	5.8	0.03	19.9	1128
LM (pond 2)	0.86	#58.1		0.01	45.5	3979
MM (pond 4)	0.99	#71.6		0.06	19.1	1249
HM (pond 3)	0.98	48.3	14.5	0.06	32.5	817
Season 2						
NM (pond 1)	0.93	33.5	17.8	0.06	255.3	583
LM (pond 2)	0.58	50.0	144.0	0.01	0.2	5000
LM (pond 3)	0.85	#168.4		0.11	74.0	1501
LM (pond 4)	0.13	10.5	58132.0	0.04	5236.0	259

P^B_{max} represents maximum intensity of photosynthesis, α the - light utilization coefficient expresses as mg C h^{-1} (mg Chl a)$^{-1}$ (μ E m^{-2} h^{-1})$^{-1}$ and I_k- light saturation μ E m^{-2} s^{-1} and P^B photosynthetic intensity expressed in mg C (mg Chl a)$^{-1}$ h^{-1} the Values represent means for monthly computations in S1 and S2. # indicates P^B_{max} that did not reach light saturation values given are for maximum attained from the curves. * denotes values that were too low.

Table 3.6a. Taxonomic composition of phytoplankton in Gaba Fingerponds, Seasons 1 and 2.

Group	Manure level (pond no)	Season 1				Season 2			
		NM (3)	NM (4)	LM (1)	MM (2)	NM (4)	LM (1)	LM (2)	LM (3)
Bacillariophyta	Aulacoseria spp.	1 (+)	2 (+)	3 (+)	2 (+)	1 (+)	1 (+)	1 (+)	1 (+)
	Cyclostephanos sp.	1 (+)	1 (+)	1 (+)	1 (+)	1 (+)	1 (+)		1 (+)
	Cyclotella sp.	1 (+)		1 (+)	1 (+)				1 (+)
	Epithemia spp.	1 (+)	1 (+)	1 (+)			1 (+)		1 (+)
	Gomphonema sp.								
	Navicula spp.	1 (+)	2 (+)	2 (+)		1 (+)	1 (+)		2 (+)
	Nitzschia spp.	3 (+)	2 (+)	2 (+)	2 (2)	2 (2)	1 (+)	2 (+)	2 (+)
	Rhopalodia spp.					2 (+)	1 (+)	2 (+)	
	Stephanodiscus sp.		1 (+)						
	Surirella sp.	1 (+)	1 (+)						
	Synedra sp.			1 (+)	1 (+)			1 (+)	
Chlorophyta	Ankistrodesmus spp.	3 (5)	3 (3)	3 (+)	2 (4)	2 (3)	3 (5)	2 (3)	2 (3)
	Characium sp.	1 (+)		1 (+)					
	Chodatella sp.	1 (+)	1 (+)	1 (+)					
	Closterium spp.	1 (+)	1 (+)	2 (+)	2 (+)	1 (+)	1 (+)	1 (+)	
	Coelastrum sp.	1 (+)		1 (+)					
	Cosmarium sp.		1 (+)						
	Crucigenia spp.	2 (3)	3 (+)	3 (3)	1 (+)	1 (+)	1 (+)	2 (+)	1 (+)
	Golenkia sp.			1 (+)					
	Kirchneriella sp.							1 (+)	
	Monoraphidium spp.	2 (+)	1 (+)	1 (+)	4 (+)	1 (+)	1 (2)	1 (+)	1 (+)
	Oocystis spp.	2 (+)	1 (+)	2 (+)	1 (+)	1 (+)	1 (+)	1 (+)	2 (+)
	Pediastrum spp.	3 (+)	1 (+)	2 (+)	1 (+)	1 (+)			
	Scenedesmus spp.	3 (+)	4 (+)	4 (+)	2 (+)	3 (+)	3 (+)	3 (+)	3 (+)
	Stuarastrum spp.	1 (+)	1 (+)	1 (+)			1 (+)		
	Tetraedron spp.	2 (2)	2 (+)	2 (+)	2 (+)	1 (+)	1 (+)	1 (+)	1 (2)
Cryptophyta	Cryptomonas spp.	2 (+)	4 (+)	4 (2)	2 (+)	1 (+)	1 (+)	2 (3)	1 (+)
	Rhodomonas spp.	2 (2)	2 (+)	2 (+)	1 (+)	1 (+)	2 (+)	1 (+)	1 (+)
Cyanophyta	Anabaena spp.	1 (+)	3 (+)	1 (+)	2 (+)	2 (3)	2 (5)	2 (2)	2 (3)
	Anabaenopsis spp.				1 (+)				
	Aphanocapsa spp.	1 (5)	1 (4)	4 (2)	3 (4)	1 (+)	2 (4)	2 (3)	1 (+)
	Aphanothece sp.								
	Coelomoron sp.				1 (+)	1 (+)			
	Chroococcus spp.	2 (+)	2 (+)	2 (+)	2 (+)	3 (+)	2 (2)	3 (+)	1 (2)
	Coelosphaerium sp.	1 (+)	1 (+)		1 (+)		1 (+)	1 (+)	1 (+)
	Cyanodictyon sp.		1 (+)		1 (+)	1 (+)	1 (+)	1 (+)	
	Cylindrospermopsis spp.	1 (+)		2 (+)	1 (+)				
	Gomphosphaeria sp.			1 (+)					
	Merismopedia spp.	1 (+)	2 (3)	6 (3)	2 (3)	1 (3)	3 (3)	1 (2)	1 (3)
	Microcystis spp.	2 (2)	5 (2)		4 (+)	3 (3)	2 (2)	3 (2)	3 (5)
	Picocyanobacteria sp.	1 (+)	1 (+)	1 (5)	1 (5)		1 (+)		
	Planktolynbya spp.	2 (+)	3 (+)	3 (+)	3 (2)	4 (3)	3 (4)	4 (3)	2 (5)
	Pseudoanabaena spp.								1 (+)
	Romeria spp.	1 (+)	2 (+)	2 (+)	1 (+)	1 (+)	1 (+)	1 (2)	1 (+)
Dinophyta	Glenodinium sp.			1 (+)				1 (+)	1 (+)
	Peridinium spp.	1 (+)	2 (+)		1 (2)				1 (+)
Euglenophyta	Colacium sp.	1 (+)							
	Euglena spp.	1 (2)	1 (2)	2 (+)	2 (2)	1 (2)	1 (+)	1 (3)	1 (4)
	Phacus spp.	2 (2)	1 (+)	1 (+)	2 (+)			1 (+)	1 (+)
	Strombomonas spp.			1 (+)	1 (+)				1 (+)
	Trachelomonas spp.	1 (2)	3 (2)	3 (2)	4 (3)	1 (2)	1 (+)	1 (+)	1 (2)
Total no of spp.		**54**	**63**	**70**	**59**	**39**	**41**	**43**	**39**

Samples taken monthly from the deep end of the ponds. Values presented are maximum number of species present and in parenthesis the percentage cell density ranked as follows: + denotes < 10% (rare), 2: 11-30% (occasional), 3: 31-50% (frequent), 4: 31-60 % (abundant) and 5: > 60 % (dominant) modified from the DAFOR scale factor. Species underlined were present frequently to dominantly at any one time and values are bolded.

Table 3.6b. Taxonomic composition of phytoplankton in Walukuba Fingerponds, Seasons 1 and 2.

Group	Manure level (pond no)	Season 1				Season 2			
		NM (1)	LM (2)	MM (4)	HM (3)	NM (1)	LM (2)	LM (3)	LM (4)
Bacillariophyta	*Aulacoseria spp.*	3 (+)	1 (+)	2 (+)	2 (+)	2 (+)	2 (+)	2 (+)	1 (+)
	Cyclostephanos sp.	1 (+)	1 (+)	1 (+)	1 (+)	1 (+)	1 (+)	1 (+)	
	Cyclotella sp.	1 (+)		1 (+)		1 (+)	1 (+)	1 (+)	
	Epithemia spp.		2 (+)		1 (+)	1 (+)			2 (+)
	Gomphonema sp.					1 (+)			
	Navicula spp.		2 (+)	3 (+)	2 (+)	2 (+)	1 (+)	1 (+)	
	Nitzschia spp.	3 (3)	3 (3)	3 (3)	3 (5)	3 (2)	3 (+)	3 (+)	3 (3)
	Rhopalodia spp.			1 (+)		1 (+)			
	Stephanodiscus sp.	1 (4)			1 (+)				
	Surirella sp.					1 (+)			1 (+)
	Synedra sp.	1 (+)	1 (+)	1 (+)	1 (+)	1 (+)	1 (+)	1 (+)	1 (+)
Chlorophyta	*Ankistrodesmus spp.*	2 (5)	1 (5)	2 (4)	2 (5)	3 (2)	3 (5)	3 (2)	2 (2)
	Characium sp.								1 (+)
	Chodatella sp.				1 (+)	1 (+)	1 (+)		1 (+)
	Closterium spp.	1 (+)	1 (+)	1 (+)	1 (+)	1 (+)	1 (+)	1 (+)	1 (+)
	Coelastrum sp.					1 (+)			
	Cosmarium sp.				1 (+)		1 (+)		
	Crucigenia spp.	2 (5)	2 (2)	1 (+)	1 (2)	3 (+)	2 (+)	2 (+)	3 (2)
	Golenkia sp.								
	Kirchneriella sp.					2 (+)	1 (+)	1 (+)	1 (+)
	Monoraphidium spp.	1 (+)	2 (+)	1 (+)	1 (+)	3 (+)	3 (2)	2 (+)	2 (+)
	Oocystis spp.	1 (+)	1 (+)	2 (+)	1 (+)	1 (+)	2 (+)	2 (+)	1 (+)
	Pediastrum spp.	1 (+)		2 (+)	1 (+)	2 (+)	1 (+)	1 (+)	1 (+)
	Scenedesmus spp.	5 (4)	6 (2)	6 (+)	5 (+)	4 (+)	6 (+)	4 (+)	5 (+)
	Stuarastrum spp.		1 (+)	1 (+)	1 (+)	2 (+)		1 (+)	2 (+)
	Tetraedron spp.	1 (+)	1 (+)	1 (+)		1 (+)	1 (+)	1 (+)	
Cryptophyta	*Cryptomonas spp.*	1 (2)	1 (+)	1 (+)	1 (+)	2 (+)	2 (+)	1 (+)	3 (5)
	Rhodomonas spp.	1 (2)	1 (2)	1 (+)	1 (+)	1 (+)	1 (+)	1 (3)	1 (2)
Cyanophyta	*Anabaena spp.*	2 (+)	2 (+)	2 (+)	2 (+)	3 (+)	2 (+)	2 (+)	3 (+)
	Anabaenopsis spp.					1 (+)	1 (+)	2 (+)	1 (+)
	Aphanocapsa spp.	2 (5)	2 (2)	2 (2)	2 (2)	3 (2)	3 (3)	3 (3)	3 (3)
	Aphanothece sp.				1 (+)		1 (+)		1 (+)
	Coelomoron sp.								1 (+)
	Chroococcus spp.	2 (+)	2 (+)	2 (3)	2 (+)	3 (+)	3 (+)	3 (+)	3 (4)
	Coelosphaerium sp.	1 (+)	1 (+)	1 (+)	1 (+)	1 (+)			1 (+)
	Cyanodictyon sp.								1 (+)
	Cylindrospermopsis spp.	1 (+)	2 (+)	2 (+)	1 (+)	2 (+)		2 (+)	2 (+)
	Gomphosphaeria sp.								
	Merismopedia spp.	1 (2)	1 (3)	1 (3)	1 (5)	3 (5)	3 (5)	2 (3)	3 (2)
	Microcystis spp.	5 (5)	4 (+)	3 (2)	3 (+)	3 (5)	4 (+)	3 (3)	4 (+)
	Picocyanobacteria sp.					1 (5)	1 (3)		
	Planktolynbya spp.	3 (3)	3 (3)	2 (3)	3 (3)	6 (2)	3 (5)	4 (5)	5 (5)
	Pseudoanabaena spp.		1 (+)		1 (+)				2 (+)
	Romeria spp.	2 (+)		1 (+)	2 (+)	1 (+)	1 (+)	2 (+)	1 (+)
Dinophyta	*Glenodinium sp.*					1 (+)			
	Peridinium spp.		1 (+)	1 (+)	1 (5)	1 (+)	1 (+)		1 (+)
Euglenophyta	*Colacium sp.*					1 (+)	1 (+)		
	Euglena spp.	3 (+)	3 (+)	3 (+)	3 (5)	4 (+)	2 (+)	2 (+)	
	Phacus spp.	2 (+)	1 (+)	2 (+)		2 (+)	2 (+)		2 (+)
	Strombomonas spp.					2 (+)		2 (+)	2 (+)
	Trachelomonas spp.	2 (+)	4 (+)	2 (+)	3 (+)	1 (+)		2 (+)	2 (+)
Total no of spp.		52	54	55	54	80	62	60	70

Samples taken monthly from the deep end of the ponds. Values presented are maximum number of species present and in parenthesis the percentage cell density ranked as follows: + denotes < 10% (rare), 2: 11-30% (occasional), 3: 31-50% (frequent), 4: 31-60 % (abundant) and 5: > 60 % (dominant) modified from the DAFOR scale factor. Species underlined occurred frequently to dominantly at any one time and values are bolded.

Nutrient release experiment and subsequent effects on oxygen

Descriptive statistics for pH, EC, DO, temperature, NH_4-N and SRP obtained during this experiment are shown in Table 3.7. Differences were noted for some parameters within and between locations. EC was notably different between locations and between ponds while pH and temperature did not show great variation with pH ranging from 7.4 to 9.5 while temperatures ranged from 22.4 to 30.3 °C. Higher values of pH and EC were attained in Walukuba and more notably for EC which ranged from 1038 to 1828 μS cm^{-1} in Walukuba compared to 223 to 530 μS cm^{-1} in Gaba. DO concentrations decreased away from the manure application cages in the ponds with exception of pond 2 (LM) in Walukuba (Figure 3.9). Over the day higher variations in DO concentrations occurred in LM and HM ponds in Walukuba and were more pronounced in the HM pond.

Concentrations of ammonia (NH_4-N) in the water column did not increase significantly. In comparison, soluble reactive phosphorus (SRP) increased significantly in all ponds with exception of the MM pond (pond 4) in Walukuba (Figure 3.10). In general SRP concentrations where higher within 3 days. A logarithmic relationship was found between the nutrient concentrations (NH_4-N & SRP) and time with SRP having a stronger correlation. Unmanured ponds had lower nutrient concentrations with no major effects of distance or time and are thus not graphically presented.

Regression results for effects of location, distance, time and manure input on DO, NH_4-N and SRP are shown in Table 3.8 (Models 1-3). The models were all significant (P<0.001) for corresponding 1303, 264 and 262 cases. Four independent variables were used: time of day, time in days, distance from manure application cage, location (Gaba) and manure load. Time of day (i.e. at 06.00 hours was recognized as attaining the lowest DO while at 18.00 hours values were higher) was the most important explanatory variable having a negative effect on DO (Table 3.8, Model 1). Time in days, location (Gaba) and manure load also had a significant negative effect but with low beta weights were not considered important. In model 2, NH_4-N was significantly and positively affected by manure input and location with the former as the most important explanatory variable. Model 3 showed that SRP was positively and significantly affected by time in days and manure input. SRP was higher in Walukuba than in Gaba. Time in days was ranked as the most important explanatory variable though in this case all beta weights were less than 0.500.

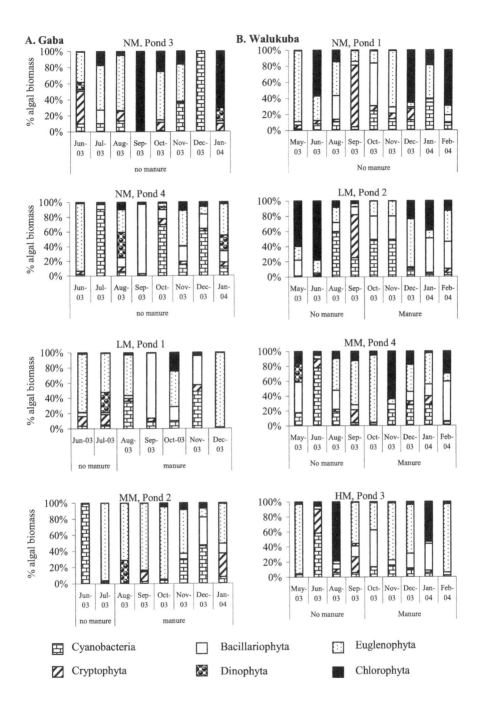

Figure 3.7. Temporal variation in algal taxonomic composition in Gaba and Walukuba, season 1 (manuring period only). Integrated samples collected monthly from ponds deep ends.

70 Chapter 3

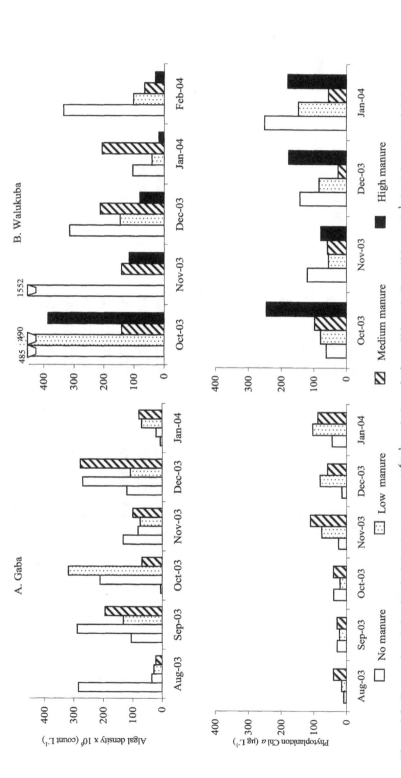

Figure 3.8. Temporal variation in algal density (count × 10⁶ L⁻¹) and phytoplankton Chlorophyll *a* (Chl *a* µg L⁻¹) in Gaba and Walukuba. Season 1 (manuring period). Integrated samples collected monthly from deep end. Bars with numbers are for values not fitting the scale.

Table 3.7. Water quality results for the nutrient release experiment.

(a) Gaba

Parameter	No manure (Ponds 3 & 4) (n = 33)			Low Manure (n = 66)			Medium manure (n = 66)		
	Min	Max	Mean (SE)	Min	Max	Mean (SE)	Min	Max	Mean (SE)
pH	7.9	8.6	8.3 (0.00)	7.9	9.1	8.7 (0.00)	7.4	8.8	8.2 (0.04)
EC (μS cm^{-1})	469	530	501 (3.60)	435	497	163 (2.20)	332	379	352 (1.50)
DO (mg L^{-1})	4.1	5.9	5.2 (0.08)	4.5	9.8	6.4 (0.14)	3.7	7.6	5.4 (0.14)
Temperature ($^{\circ}$C)	23.5	27.4	26 (0.20)	23.2	28.2	25.9 (0.20)	22.4	28.1	25.0 (0.20)
NH$_4$-N (mg L^{-1})	0.110	0.350	0.190 (0.01)	0.190	0.042	0.260 (0.01)	0.150	0.420	0.300 (0.01)
SRP (mg L^{-1})	0.000	0.002	0.010 (0.00)	0.004	0.110	0.040 (0.00)	0.060	0.210	0.090 (0.03)

(b) Walukuba

Parameter	No manure (Pond 1) (n = 15)			Low Manure (n = 30)			Medium manure (n = 30)			High manure (n = 30)		
	Min	Max	Mean (SE)	Min	Max	Mean (SE)	Min	Max	Mean (SE)	Min	Max	Mean (SE)
pH	8.5	8.9	8.7 (0.03)	8.8	9.1	8.9 (0.02)	8.2	9.3	9.2 (0.00)	9.0	9.5	9.3 (0.00)
EC (μS cm^{-1})	1038	1170	1105 (11.90)	1215	1477	1329 (15.30)	1556	1828	1643 (16.00)	1460	1743	1592 (15.00)
DO (mg L^{-1})	4.9	9.7	8.2 (0.34)	3.0	10.9	7.0 (0.47)	6.0	10.1	8.1 (0.24)	3.9	7.4	11.7 (0.71)
Temperature ($^{\circ}$C)	24.4	29.7	27.2 (0.59)	23.9	30	27.2 (0.40)	24.9	31.9	28.0 (0.43)	23.9	30.3	27.2 (0.41)
NH$_4$-N (mg L^{-1})	0.010	0.200	0.120 (0.02)	0.010	0.250	0.160 (0.01)	0.020	0.250	0.15 (0.01)	0.020	0.300	0.200 (0.07)
SRP (mg L^{-1})	0.001	0.645	0.091 (0.05)	0.030	0.560	0.260 (0.03)	0.070	0.900	0.15 (0.03)	0.010	0.870	0.390 (0.06)

Values represented are minimum and maximum (represented as min and max), means and in parenthesis standard errors of the mean denoted as SE and number of samples denoted as n.

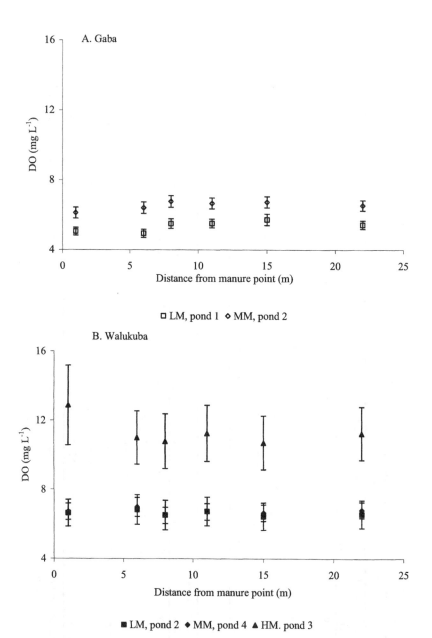

Figure 3.9. Variation in dissolved oxygen (DO) over pond length following manure application (the first 30 days). Values plotted are means for measurements taken at 06.00, 12.00, 14.00 and 18.00 hours daily for A. Gaba; the first 8 days, every three days in the second week and weekly from day 16 and B. Walukuba; days 1, 3, 7, 14 and 21. Error bars represent standard error of the mean and show variation in time.

A. Gaba

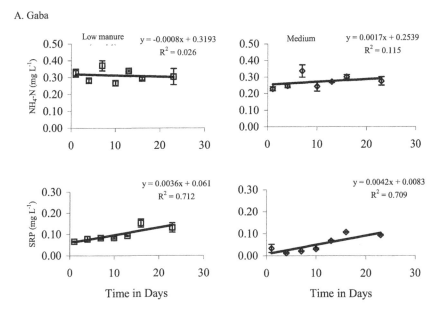

Figure 3.10a. Changes in ammonium nitrogen (NH₄-N) and soluble reactive phosphorus (SRP) following manure application in Gaba. Values plotted are means for nutrient concentrations at different points in the pond at 1, 8, 15 and 22 m from the manure cage. Bars represent standard error of the mean and indicate spatial variation.

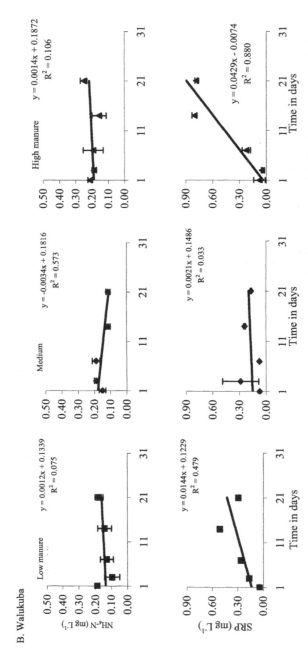

Figure 3.10b. Changes in ammonium nitrogen (NH_4-N) and soluble reactive phosphorus (SRP) following manure application in Gaba. Values plotted are means for nutrient concentrations at different points in the pond at 1, 8, 15 and 22 m from the manure cage. Bars represent standard error of the mean and indicate spatial variation.

Table 3.8. Multiple regression models for dissolved oxygen (mg L^{-1}) with independent variables: distance from the manure point (m), dummy location, manure load (kg ha^{-1} (2weeks^{-1}), time of day (06.00hrs) and times in days (days)

	Model 1 - LNDO (N = 1303)					Model 2 - LNNH$_4$-N (N = 264)					Model 3 - LNSRP (N = 264)				
	B	SE	Beta	Sig.	Rank	B	SE	Beta	Sig.	Rank	B	SE	Beta	Sig.	Rank
Distance from manure point (m)	0.001	0.001	0.009		5	0.000	0.000	-0.043		4	-0.002	0.008	-0.013		4
Dummy location	-0.238	0.020	-0.239	***	2	0.098	0.007	0.713	***	1	-0.979	0.131	-0.354	***	2
Manure load (kg ha^{-1} (2 weeks)$^{-1}$)	0.013	0.001	0.212	***	3	0.003	0.000	0.349	***	2	0.047	0.008	0.287	***	3
Time of day (0600 hrs)	0.258	0.008	0.592	***	1										
Time in days (days)	-0.004	0.001	-0.059	**	4	0.001	0.000	0.056		3	0.087	0.009	0.441	***	1
Constant	1.219					0.103					3.276				
Adjusted R^2	0.507					0.453					0.481				
F value	269.071					53.371					61.412				
Probability	<0.001					<0.001					<0.001				
Durbin-Watson Statistic	0.332					1.502					0.554				

Significant levels for partial regression coefficients are marked at * 5%, ** at 1% and *** at 0.1 %, number of cases = 1,303, 264 and 254 for models 1, 2 and 3.

Discussion

Effect of pond inputs/conditions on water quality

Water quality in the Fingerponds was influenced by differences between location and seasonal effects. The pH values of 6.9 to 9.9 and temperatures (between 20 and 30°C) are normal for fishponds (Stone & Thomforde, 1977; Boyd, 1997; Hargreaves & Heusel, 2000). Though most fish favour waters of pH 6.5-9.0, for tilapia, a temperature of 28 to 32°C is recommended (Teichert-Coddington et al., 1997).

Variation in water quality can be attributed to a number of factors, including mineral load (Golterman & Kouwe, 1980). In this study, conductivity (EC) values differed notably between the locations and between ponds, with higher values (above 1000 µS cm^{-1}) obtained in Walukuba. This was probably due to differences in soil characteristics and underlying geological formations, as well as seasonality effects on EC. With low water levels during the second season, minerals became concentrated in the ponds but EC did not reach 5000 µS cm^{-1}; a level that can affect the osmo-regulation of fish (Stone & Thomforde, 1977).

In both locations, alkalinity was high and exceeded 200 mg CaCO$_3$ L^{-1} which could have reduced carbon fixation in photosynthesis at the highest values. Photosynthesis was further reduced due to light limitation arising from the high turbidity and suspended solids (TSS) which occurred in both locations. In Gaba which registered significantly higher levels, a slight reduction was observed in the manured ponds. This could be an initial improvement of sedimentation at the onset of manuring (Ryding & Rast, 1989) but since TSS increased thereafter it appears transitory. With falling water levels over the seasons, pond transparencies became lower and Chl a concentrations tended to decline. The high turbidity in Walukuba mainly arose from phytoplankton Chl a as shown by the reduced effects of high TSS on Chl a values. On the contrary in Gaba, the high turbidity resulted from clay particles eroded from pond edges and stirred up the bioturbation by fish and ostracods as previously pointed out by Porkornỳ et al. (2005).

Oxygen dynamics in semi-intensive aquacultural systems are generally dominated by photosynthesis and respiration of phytoplankton populations. The Fingerponds followed the typical daily oxygen patterns of fish ponds (Diana et al., 1991): DO values were always greater than 2 mg L^{-1} during the day, and apart from the highly manured pond in Walukuba, no distinct differences in oxygen concentrations were noted between ponds with different manure input. This is contrary to findings by other authors (Melack, 1979; Wetzel, 2001). In Gaba, this disparity may be attributed to the high density of Pico-Cyanobacteria in the manured ponds which resulted in diffusion of oxygen into the air due to their ability to form thick floating blooms (Delincé, 1992).

Manure inputs in the ponds resulted in distinct differences in the concentrations of NH$_4$-N and SRP. Total phosphorus concentrations exceeded 100 µg L^{-1} rising to concentrations over 1.0 mg L^{-1} (especially at Walukuba) and were higher than many other fishpond systems (Boyd, 1990; Diana et al., 1997). The high values are indicative of a hypertrophic state (Wetzel, 2001). NO$_3$-N concentrations in the ponds were generally low (< 200 µg L^{-1}) with occasional highs of up to 900 µg L^{-1} in the LM pond in Gaba (Season 1). The TN concentrations were variable and distinctively increased with manure input. Like other fishpond systems (McNabb et al., 1990; Knud-Hansen et al., 1991) the N:P ratios (< 7) were generally low indicating the likelihood of N limitation probably influenced by the phosphorus-rich soils in the region (Henao & Baanante, 1999). Low water levels, particularly in the second season, resulted in higher nutrient concentrations.

Ammonia is potentially toxic to fish particularly in high pH waters (Boyd, 1990). Following application of manure, NH_4-N was released slowly into the water column primarily through leaching and breakdown of soluble organic matter (Knud-Hansen et al., 1991; Nath & Lannan, 1993). Despite this slow release, high NH_4-N concentrations were detected in ponds and could have arisen from increased sedimentation/re-suspension due to increased organic loading, phytoplankton death and fish excreta (Hargreaves, 1998; Hargreaves & Tucker, 2004). Diel fluctuations of NH_4-N were similar to other fish ponds with the highest concentrations observed after sunrise and a general decrease during the day due to algal uptake (Shilo & Rimon, 1982; Abdalla, et al., 1996).

Phytoplankton production and primary productivity

Increased dominance of Cyanobacteria was observed in the ponds after manure application. Commonly found were species such as *Microcystis* and *Merismopedia*. These taxa have the affinity to absorb N and P at severely limiting levels (Harding, 1997; Schagerl & Oduor, 2003) and can thrive in waters with low N: P ratios (Rhee & Gotham, 1980; Watson et al., 1997) and low NO_3-N concentrations (Hargreaves, 1998). Low N:P ratios further encouraged the presence of nitrogen fixing Cyanobacteria such as *Anabaena* spp. but discouraged green algae as reported by Smith (1983). The presence of Cryptophyta was probably due to the high turbidities favouring their growth (Watson et al., 1997). The low numbers of Bacillariophyta in the Fingerponds are attributed to low NO_3-N concentrations (Lomas & Glibert, 1999) and water turbidity that can hinder their growth and lead to rapid sedimentation of available diatoms (Reynolds, 1984).

A cyclic pattern for both biomass and abundance of phytoplankton, demonstrated by the monthly values, does not elucidate fully the phytoplankton dynamics due to the long period between sampling dates (i.e. 4 weeks). As shown by the regression models, phytoplankton Chl *a* concentrations were much higher in the second season which is related to the lowering of pond water levels and resultant concentration of nutrients for algal growth. The high turbidities in Gaba generally limited biomass production. Although high phytoplankton biomasses occasionally occurred in the unmanured ponds in the first season through Chlorophyta proliferation, N limitation and light attenuation hindered phytoplankton production, further aggravated by decreasing water levels particularly in Gaba.

Net primary productivity (NPP) was enhanced with manure application with some differences noted between ponds but, because of falling water levels in the second season, it was not possible to optimize application rates. NPP (mean ± SE) in manured ponds was 2.9 ± 0.65 g C m^{-2} d^{-1} in Gaba (SE representing standard error of the mean, n = 12) and 2.6 ± 0.34 g C m^{-2} d^{-1} in Walukuba (n = 15) (conversion of oxygen to C based on Strickland, 1960). These values are higher than findings in Honduras (2.4 g C m^{-2} d^{-1}) but lower than values in Panama (4.4 g C m^{-2} d^{-1}) for organically fertilized ponds (Batterson et al., 1989; Green et al., 1989; Knud-Hansen et al., 1991; Teichert-Coddington et al., 1992). Primary productivity took place mainly in the upper 30 cm layer, the high clay turbidity, particularly in Gaba, suppressing it below this depth.

Phytoplankton photosynthetic intensities varied between seasons and locations showing the differences in light effects on photosynthesis in the ponds. Low P^B_{max} were observed in ponds with low light availability (high turbidity) and high Chl a more so in the second period. This is a result of the lower pond water levels which resulted in nutrients concentrating in the ponds and more re-suspension of sediment. However the differences between manured and unmanured ponds were not so distinct. Some manured ponds (e.g. the LM and MM ponds in Season 1 and LM pond 3 in Season 2) failed to attain light saturation: with dense algal populations (Chl $a > 250$ μg L^{-1}), self-shading is probable.

Factors affecting responses to manure inputs

Clay turbidity is the most prevailing problem for fertilization management in freshwater ponds (Lin *et al.*, 1997). In this study, high turbidity levels resulted in low pond transparency whilst the colloidal particles tended to adsorb mineral nutrients making them less available for uptake by the algae.

In Walukuba light attenuation in the ponds mainly arose from increased algal biomass; a phenomenon that has been reported by other researchers (Teichert-Coddington *et al.*, 1992; Diana *et al.*, 1997). High extinction coefficients (6.5-16.2) especially through clay suspensions were a problem in our Fingerponds; elsewhere values > 6 have been shown to be a major limitation to production (Yusoff & McNabb, 1989). The data confirm this as less turbidity produced higher NPP, hence the negative correlation between GPP and turbidity ($r^2 = -0.38$, p< 0.001). In addition, turbidity reduced the daily fluctuations of oxygen, clearly demonstrated by the diel measurements in Gaba with clay suspensions. In Walukuba with more turbidity due to algae, higher oxygen values were attained during the day.

Interaction of light and temperature further determine photosynthetic activity and the dynamics of fish ponds. Although surface photo-inhibition effects may have occurred in the ponds, the rapid light attenuation would have reduced the severity of enzymatic damage thus encouraging high productivities in the top 10 cm. Intense solar radiation will cause thermal stratification in fish ponds (Delincé, 1992) and an anoxic hypolimnion can develop. Towards morning, as the pond approaches homothermia, mixing can be triggered and a sudden drop in overall oxygen concentration may lead to fish kills. In this study, relatively low nitrogen and N: P ratios were found in nearly all ponds. Hargreaves (1998) emphasized the problem of limitation of primary productivity due to low nitrogen concentrations in fish ponds where fish yields are dependent on the development of autotrophic food webs (hence the application of organic or inorganic fertilizers). The moisture and nutrient content of manures may vary considerably due to the storage prior to use (Boyd, 1997; Lin *et al.*, 1997). In this study, the nutrient content of the chicken manure was lower than reported by other researchers (Tacon, 1987; Lin *et al.*, 1997) which may explain the rather low nitrogen in the pond water and the overall lower than expected primary productivity in manured ponds.

Finally, a reduced water level is a critical factor in primary productivity limitation in these ponds, especially in Gaba, where it is demonstrated as a key contributor to water quality variability in the regression models (water level strongly correlated with secchi depth and season). Reducing pond water levels coupled with high algal densities increased light limitation and ultimately reduced phytoplankton productivity. This was most apparent in the second season where low water levels prevailed.

Management

Fertilization of ponds increased pond productivity but a number of factors influence the effectiveness of the manure applied. High manure levels applied in Walukuba did not result in high primary productivity because of self-shading. Pond volume had an important bearing on the fertilization regimes: as water levels dropped, particularly during the second season, so did the primary productivity. Fertilization regimes in Fingerponds are therefore governed to a large extent by their water levels and must be regulated accordingly. From this study, application of chicken manure fortnightly resulted in pond environmental factors being kept within acceptable limits. Oxygen depletion was not a problem during the two experimental seasons and no fish mortalities occurred. However, as the ponds age and thick organic layers develop at the pond bottom, oxygen depletion is likely to become a problem (Boyd et al., 2002). Removal of pond soils at the end of a grow-out season is therefore necessary and can be used as a useful fertilizer in horticulture.

Ammonia concentration is a major factor that controls fish production in fish farming systems. Un-ionised ammonia (UIA) which is the toxic form of ammonia may limit fish production if the ammonia assimilation capacity of the pond is exceeded (El-Shafai, 2004). In fish ponds, the toxic form of ammonia is likely to predominate in the late afternoon and early evening when pH values are highest but since this period is not prolonged an exertion on the oxygen demand is unlikely to occur. Although the accumulation of NH_3 to toxic levels in ponds is unlikely (Abdalla, 1989), this threat cannot be ruled out. Toxic levels for UIA for short exposure lie between 0.6 and 2 mg L^{-1} with maximum tolerable concentrations at 0.1 mg L^{-1} (Pillay, 1992). Algal crashes can cause dramatic increments in ammonia levels in ponds up to 6-8 mg L^{-1} with subsequent drop in pH (7.8-8.0) (Hargreaves & Tucker, 2004). In this study, NH_4-N concentrations rose to 9.9 mg L^{-1} in Season 2 at a pH of 8.4 corresponding to 1.2 mg L^{-1} NH_3 (computed according to Meade, 1985) which is still within the safe ranges for short exposure. Ammonia levels in the Fingerponds are best controlled by regulation of manure inputs cognizant of dropping water levels. Finally, stocking densities play an important role in safeguarding the system against ammonia toxicity. Periodic harvesting of fish over the grow-out period could help maintain optimum fish densities.

Conclusions

Nutrient levels and primary productivity in the Fingerponds were enhanced with manure application, but not to the desired levels. This was due to a number of limitations: low N: P ratios, high turbidity and decreasing water levels. Differences in location (the pond hydrological systems) and seasonal effects played a role in the variability of the pond water quality thus influencing the effectiveness of the manure input/conditions. Turbidity sources varied for the two locations with turbidity in Gaba mainly due to inorganic clay while in Walukuba it was due to phytoplankton densities. Oxygen depletion and ammonia toxic effects are a constant threat to Fingerpond systems particularly with falling water levels over the grow-out period. Water quality is best regulated by careful and accurate fertilization regimes.

Application of chicken manure (520-1,563 kg ha^{-1} pond^{-1} fortnightly) was adequate to maintain a reasonable level of productivity in the ponds. Care must be taken to use good quality manure. Phytoplankton development in the ponds was regulated mainly by the nutrient levels and resulted in dominance of Cyanobacteria. Water quality in the Fingerponds over the two seasons did not deteriorate to levels unsuitable for fish but may have adverse effects on fish productivity.

Acknowledgments

This research was funded by the European Union, EU/INCODEV Fingerponds Project No. ICA4–CT–2001–10037 and in part by the Netherlands Government through the Netherlands Fellowship Programme through UNESCO–IHE, Institute for Water Education, Delft, The Netherlands. It was carried out in partnership with Makerere University Institute for Environment and Natural Resources (MUIENR). We are indebted to the management and staff of National Water and Sewerage Corporation (NWSC), Uganda for providing technical support and facilities. We thank all the field and laboratory staff who worked tirelessly.

References

Abdalla, A.A.F. 1989. The effect of ammonia on *Oreochromis niloticus* (Nile Tilapia) and its dynamics in fertilized tropical fish ponds. PhD dissertation. Michigan State University, East Lansing. 62 pp.

Abdalla, A.A.F., McNabb, C.D. and Batterson, T.R. 1996. Ammonia dynamics in Fertilized Fish Ponds Stocked with Nile Tilapia. *The Progressive Fish-Culturist* 58: 117-123.

APHA, 1992. *Standard methods for examination of water and wastewater.* 18th Edition. American Public Health Association. Washington, DC.

APHA, 1995. *Standard methods for examination of water and wastewater,* 19th Edition. American Public Health Association, Washington, DC.

Batterson, T.R., McNabb, C.D., Knud-Hansen, C.R., Eidman, H.M. and Sumatadinata, K. 1989. Indonesia: Cycle III of the Global Experiment In *Pond Dynamics/Aquaculture CRSP* Data Report, Vol. 3, No. Egna, H.S.(ed.), Oregon State University, Corvallis, 135 pp.

Boyd, C.E. 1990. *Water quality in ponds for aquaculture.* Birmingham Publishing Co., Birmingham, Alabama. 482 pp.

Boyd, C.E. 1997. Practical aspects of chemistry in pond aquaculture. *The Progressive Fish-Culturist* 59: 85-93.

Boyd, C.E., C.W. Wood, and T. Thunjai. 2002. *Aquaculture pond bottom soil management.* USAID PD/A CRSP. Oregon State Univ., Corvallis, OR. 41 pp.

Carpenter, S.R., Kitchell, J.F. and Hodgson. J.R. 1985. Cascading trophic interactions and lake productivity. *BioScience* 35: 634-639.

Delincé, G. 1992. *The ecology of the fish pond ecosystem with special reference to Africa.* Kluwer Academic Publishers, Dordrecht. 230 pp.

Denny, P., Kipkemboi, J., Kaggwa, R. and Lamtane, H. 2006. The potential of fingerpond systems to increase food production from wetlands in Africa. *International Journal of Ecology and Environmental Sciences* 32 (1): 41-47.

Diana, J.S., Lin, C.K. and Schneeberger, P.J. 1991. Relationships among nutrient inputs, water nutrient concentrations, primary production and yield of *Oreochromis niloticus* in ponds. *Aquaculture* 92: 323-343.

Diana, J.S., Szyper, J.P., Batterson, T.D., Boyd, C.E. and Piedrahita, R.H. 1997. Water quality in ponds. In: H.S. Egna & Boyd, C.E. (eds.) *Dynamics of pond aquaculture*. CRC Press, Boca Raton, New York. pp. 53-71.

El-Shafai, S.A.A.M. 2004. *Nutrients Valorization via Duckweed-based Wastewater Treatment and Aquaculture*. Ph.D Thesis. A.A. Balkema Publishers. 174 pp.

Golterman, H.L. and Kouwe, F.A. 1989. Chemical budgets and nutrient pathways. In: E.D.Le Cren & R.H. Lowe-McConnell (eds). *The functioning of freshwater ecosystems*. Cambridge University Press, Cambridge International Biological Programme. 22. pp 85-140.

Green, B.W., Phelps, R.P. and Alvarenga, H.R.1989. The effect of manures and chemical fertilizers on the production of *Oreochromis niloticus* in earthen ponds. *Aquaculture* 76: 37-42.

Harding, W.R. 1997. Phytoplankton primary production in a shallow, well-mixed, hypertrohic South African Lake. *Hydrobiologia*. 344: 87-102.

Hargreaves, J. A. 1998. Nitrogen biochemistry of aquaculture ponds. *Aquaculture* 166: 181-212.

Hargreaves, J.A. and Heusel, L. 2000. A control system to stimulate diel pH fluctuation in eutrophic aquaculture ponds. *Journal of World Aquaculture Society* 31: 390- 402.

Hargreaves, J.A. and Tucker, C.S. 2004. Managing ammonia in fish ponds. *SRAC. Publication* 4603. 7 pp.

Hecky, R.E. and Kling, H. J. 1981. The phytoplankton and protozooplankton of the euphotic zone of Lake Tanganyika: Species composition, biomass, chlorophyll *a* content and spatial-temporal distribution. *Limnology and Oceanogrphy* 26 (3): 532-547.

Hecky, R.E. and Kilham, P. 1988. Nutrient limitation of phytoplankton in fresh water and marine environments: a review of recent evidence on the effects of enrichments *Limnology and Oceanography* 33: 796-832.

Henao, J. and Baanante, C. 1999. Nutrient depletion in agricultural soils of Africa. In: Pinstrup-Andersen, P. & Pandya-Lorch, R. (eds.). The unfinished agenda. Part 6. Perspectives in overcoming hunger, poverty and environmental degradation. *International Food Policy Research Institute* (IFRI). Washington, D.C. pp. 159-163.

Hopkins, K.D., Lannan, J.E. and Bowman, J.R. 1988. Managing a database for pond research data-the CRSP experience. *Aquabyte* 1. 1. pp. 3-4.

John, D.M., Whitton, B.A. and Brook, A.J. (eds). 2002. *The freshwater algal flora of the British Isles. An identification guide to freshwater and terrestrial algae*. Cambridge University press. 702 pp.

Kent, M. and Coker, P. 1992. *Vegetation description and analysis: A practical approach*. CRC, Boca Raton. 363 pp.

Knud-Hansen, C.F. 1997. Experimental design and analysis in aquaculture In: H.S. Egna & Boyd C.E. (eds). *Dynamics of pond aquaculture*. CRS Press Boca/ Raton New York. pp.325-375.

Knud-Hansen, C.F., Batterson, T.R., McNabb, C.D., Harahat, I.S., Sumantadinata, K. and Eidman, H.M. 1991. Nitrogen input, primary productivity and fish yield in fertilized freshwater ponds in Indonesia. *Aquaculture* 94: 49-63.

Lin, C.K., David, R., Teichert-Coddington, Green, B.W., and Veverica, K.L. 1997. Fertilization regimes. In: H.S. Egna and Boyd, C. E. (eds.) *Dynamics of pond aquaculture*. CRC Press LLC. Boca Raton, Florida. pp.73-104.

Lomas, M.W. and P.M. Glibert. 1999. Temperature regulation of nitrate uptake: a novel hypothesis about nitrate uptake and reduction in cool-water diatoms. *Limnology and Oceanography* 44: 556-572.

McNabb, C.D., Batterson, T.R., Premo, B.J., Knud-Hansen, C.R., Eidman, H.M., Lin, C.K., Jaiyen, K., Harrison, J.E., and Chuenpagdee, R., 1990. Managing fertilizers for fish yield in tropical ponds in Asia in R.A.J.H., Eds., The Proceedings of second Asian Fisheries Forum. *The Asian Fisheries Society*, Manila, The Philippines. 169 pp.

Meade, J.W. 1985. Allowable ammonia for fish culture. *Prog. Fish-Cult.* 47: 135- 145.

Melack, J.M. 1979. Photosynthetic rates of four tropical African fresh waters. *Fresh Water Biology* 9: 555-571.

Mosille, O.I.W. 1994. Phytoplankton species of Lake Victoria. In: Reports from the Haplochromis Ecology Survey Team (HEST). Report No. 30B. Zoologisch Laboratorium, Morphology Department, Leiden University, The Netherlands. 84 pp.

Mugidde, R. 1993. The increase in phytoplankton primary productivity and biomass in Lake Victoria, *Limnologia* 25: 846-849.

Nath, S.S. and Lannan, J.E. 1993. Dry matter nutrient relationships in manures and factors affecting nutrient availability from poultry manures. In: Egna, M. McNamara, J., Bowman, R. and Astin, N. (eds.) Tenth Annual Admin. Report, 1991-1992. *CRSP* Office of International Research and Development, Oregon State University, Corvallis, Oregon. pp. 110-119

Nauwerck, A. 1963. Die Beziehungen zwischen zooplankton und phytoplankton imsec. *Erhen.Symb. Bot. UpSal.* 17(5): 163 pp.

Parsons, T.R., Takahaski, M. and Hargrave, B. 1984. *Biological oceanographic processes*. 3rd Edition. Oxford, Pergamon Press. 330 pp.

Pechar, L., 1987. Use of an acetone: methanol mixture for the determination of extraction and spectrophotometric determination of chlorophyll-*a* in phytoplankton. *Arch. Hydrobiol. Suppl.* 78: 99-117.

Pillay, T.V.R. 1992. Water and wastewater use. In: T.V.R. Pillay (ed.) *Aquaculture and the environment*. pp. 49-55.

Platt, T., Gallegos, C.L., and Harrison, W.G. 1980. Photo inhibition of photosynthesis in natural assemblages of coastal marine phytoplankton, *Journal of Marine Research*. 38 (4): 687-701.

Prein, M., Hulata, G. and Pauly, D.(ed.). 1993. *Multivariate methods in aquaculture research: Case studies of tilapias in experimental and commercial systems*, ICLARM. Stud. Rev.20. Manila, Philippines. 221 pp.

Pokornỳ, J. Přikryl, I., Faina, R., Kansiime, F., Kaggwa, R.C., Kipkemboi, J., Kitaka, N., Denny, P., Bailey, R., Lamtane, H.A., and Mgaya, Y.D. 2005. Will fish pond management principles from the temperate zone work in tropical fingerponds. In: J. Vymazal. *Natural and Constructed Wetlands: Nutrients, Metals and Management*. Backhuys Publishers, Leiden, The Netherlands. pp. 382-399

Reynolds, C.S. 1984. *The ecology of freshwater phytoplankton*. Cambridge University Press, Cambridge, UK. 384 pp.

Rhee, G.J. and Gotham, I.J. 1980. Optimum N:P ratios and coexistence of planktonic algae. *J. Phycol.* 16: 486-489.

Robotham, P.W.J. 1990. Trophic niche overlap of fry and juveniles of *Oreochromis leucostictus* (Teleostei, Chichlidae) in the littoral zone of a tropical lake (L. Naivasha, Kenya*). Revue Hydrobiologie Tropicale* 23: 209-218.

Ryding , S.O. and Rast, W. (Eds.) 1989. *The Control of eutrophication of lakes and reservoirs. Man and the biosphere Series.* Vol. 1. UNESCO, Paris and Parthenon Publishing. Camforth. 314 pp.

Schagerl, M. & Oduor, S.O. 2003. On the limnology of Lake Baringo (Kenya): II. Pelagic Primary Production and Algal Composition of Lake Baringo, Kenya . *Hydrobiologia* 506-509: 297-303.

Schroeder, G.L., Alkon, A. and Laher, M. 1991. Nutrient flow in pond aquaculture In: Brune D. E. and Tomasso, J.R. (eds). Aquaculture and water quality. *World Aquaculture Society.* Baton Rouge, LA. pp. 489-505.

Shilo, M. and Rimon, R. 1982. Factors which affect the intensification of fish breeding in Israel. Part 2. Ammonia transformation in intense fish ponds. *Bamidgeh* 34: 101-114.

Smith, V.J. 1983. Low N to P favour dominance by blue-green taxa in Lake phytoplankton. *Science* 221: 669-671.

Smith, D.W. and Piedrahita, R.H. 1988. The relation between phytoplankton and dissolved oxygen in fish ponds. *Aquaculture* 68 (3): 249-265.

Stone, N.M. and Thomforde, H.K. 1977. Understanding your fish pond water analysis report. *University of Arkansas Coorperative Extension Service Printing Services.* 4 pp.

Strickland, J.D.H. 1960. Measuring the production of marine phytoplankton. *Bull. Fish. Res. Board. Can.,* 122: 1-172.

Tacon, A.G.J. 1987. The nutrition and feeding of farmed fish and shrimp - A technical manual, I. The essential nutrients, *FAO* Report GCP/RLA/075/ITA. Field Document 2. Brasilia, Brazil. 117 pp.

Teichert-Coddington, D.R., Green, B.W., and Phelps, R.P. 1992. Influence of site and season on water quality and tilapia production in Panama and Honduras. *Aquaculture* 105: 297-314.

Teichert-Coddington, D.R., Popma, T.J. and Lovshin, L.L. 1997. Attributes of tropical pond-cultured fish. In: H.S. Egna and Boyd, C.E. (eds) *Dynamics of pond aquaculture* CRC Press Boca Raton, New York. pp.183-198.

Trewavas, E.T. 1983. *Tilapiine Fishes of the genera Sarotherodon, Oreochromis and Danakilia.* British Museum (Natural History) London UK. 583 pp.

Viner, A.B. 1969. The chemistry of the water of Lake George, Uganda. *Verh. Int. Verein. Limno.* 17: 289-296.

Vollenweider, R.A. (Editor), Talling, J.F., and Westlake, D.F. 1974. *A manual on methods for measuring primary production in aquatic environments.* IBP Handbook No. 12. Second Ed. Int. Biological Programme, London. Blackwell Scientific Publications., Oxford. 225p.

Watson, S., McCauley, E. and Downing, J.A. 1997. Patterns in phytoplankton taxonomic composition across temperate lakes of differing nutrient status. *Limnology and Oceanography* 42 (3): 487-495.

Wetzel, R. G. 2001. *Limnology lakes and rivers ecosystems.* 3rd edition. San Diego Academic press. 1006 pp.

Wetzel, R. G. and Likens, G. E. 1991. *Limnological analyses.* 2nd Edition. Springer-Verlag. 391 pp.

Yamane, T. 1973. Statistics: An introductory Analysis (3rd edition). New York, Harper.

Yusoff, F.M. and McNabb, C.E. 1989. Effects of nutrient availability on primary productivity and fish production in fertilized tropical ponds. *Aquaculture* 71: 303-319.

Chapter 4

The use of artificial substrates in turbid seasonal wetland fish ponds 'Fingerponds' in Uganda and their effects on water quality

Abstract

Periphyton development on artificial substrates installed in turbid seasonal wetland fish ponds ('Fingerponds') was evaluated using four locally available plants: bamboo, *Phragmites, Raphia* and papyrus. The factors underlying variation in pond water quality as well as fish grazing effects were determined. Substrates (1.0 m^2 mats) were installed in three Fingerponds (each 24 x 8 m) equivalent to 60 and 64 m^2 of the pond surface area for periphyton colonization in two separate experiments. A third experiment was run with enclosures that contained 0.25 m^2 mats totaling to 3.1 m^2 of the pond surface area. Bamboo was superior for periphyton attachment and growth, followed by *Phragmites*. Periphyton density on mats grazed by fish was highest on bamboo (23.9 g dry matter m^{-2}) followed by *Phragmites* (18.5 g dry matter m^{-2}), papyrus (9.5 g dry matter m^{-2}) and *Raphia* (8.8 g dry matter m^{-2}) Chlorophyll *a* concentration was highest on bamboo (23.9 mg m^{-2}) followed by *Phragmites* (13.5 mg m^{-2}), papyrus (15.2 mg m^{-2}) then *Raphia* (11.3 mg m^{-2}). Periphyton ash content was lowest on bamboo (36-75 %) and highest on *Raphia* (42-87 %) while the highest ash free dry matter (AFDM) was found on bamboo (5.4 g m^{-2}). The periphyton was dominated by non-algal material as shown by the high Autotrophic Index values and its development was dynamic with taxonomic composition mainly influenced by nitrogen limitation and light availability. In general periphyton densities were low (0.2-0.7 g C m^{-2} d^{-1}) due to light limitation arising from high suspended solids and turbidity (secchi depth transparencies < 20 cm). Higher periphyton densities were obtained in manured ponds. Nevertheless, in the presence of artificial substrates, total primary productivity increased by more than 100 % in manured ponds. Pond water quality was not adversely affected. Artificial substrates have potential use in Fingerponds systems as a substrate for periphyton; a cheap source of food for fish.

Key words: artificial substrates, fingerponds, periphyton, suspended solids, turbidity, water quality.

Publication based on Chapter 4:
Kaggwa, R.C., van Dam, A.A., Kasule, D. and Kansiime, F. 2006. An initial assessment of the use of wetland plants as substrates for periphyton production in seasonal wetland fish ponds in Uganda. *International Journal of Ecology and Environmental Sciences* 32 (1): 63-74.

Introduction

Fingerponds are earthen ponds dug at the edge of natural wetlands and stocked naturally with wild fish (mainly *Oreochromis* spp.) during flooding. They are dependent on natural flood events. Fish production in these systems can be stimulated through fertilization with organic manures (Chapter 3).

In semi-intensive aquaculture systems, fish production is based on a combination of organic and chemical fertilizers and supplemental feeds. These systems are phytoplankton production units in which inorganic nutrients and sunlight generate plankton biomass harvestable by herbivorous fish (Keshavanath *et al.*, 2004). However, only 5-15 % of the nutrients put in a pond are converted to harvestable products while the rest accumulate in the pond sediment, or are discharged as effluent water to the environment, or volatilize (Edwards, 1993; Gross *et al.*, 1999; van Dam *et al.*, 2002). In response to high nutrient levels as a result of fertilization, phytoplankton blooms usually develop and may collapse, causing a sudden drop in dissolved oxygen and fish mortality (Delincé, 1992). Additionally, in aquaculture ponds the fish species that can utilize the periphyton assemblage are probably more numerous than those that are exclusively phytoplanktivorous (Dempster *et al.*, 1993; van Dam *et al.*, 2002). Periphyton systems that are based on traditional practices such as the 'acadja' fisheries in West Africa (Welcomme, 1972; Hem & Avit, 1994), and the 'brush-parks' in Malawi (Jamu *et al.*, 2003) make use of artificial substrates. Recent studies in fish ponds based on similar principles showed promising results (Azim *et al.*, 2005). For resource-poor farmers in many developing countries, this option for enhancing natural food in fish ponds is welcome since locally available substrates may reduce costs and labour demands.

Periphyton in this study refers to the complex of sessile biota attached to submerged substrata and includes algae, invertebrates, detritus and microorganisms embedded in a mucopolysaccharide matrix (Azim *et al.*, 2002; van Dam *et al.*, 2002). Artificial substrates in ponds create a 'periphyton loop' through which more nutrients are channeled to fish and the efficiency of nutrient utilization is increased. This provides an additional pathway for conversion of primary production to fish biomass (Azim 2001; van Dam *et al.*, 2002). Researchers have found tilapias growing well in these systems due to their ability to consume periphyton (Dempster *et al.*, 1993; Huchette *et al.*, 2000). Although some studies show little effect of periphyton on fish production (Shrestha & Knud-Hansen, 1994; Norberg, 1999), several others have reported enhanced fish production (Hem & Avit, 1994; van Dam *et al.*, 2002; Keshavanath *et al.*, 2002, 2004). However, periphyton-based systems have not been exploited in East Africa and are worth considering.

This study investigates the use of artificial substrates made from bamboo and local wetland plants in 'Fingerponds' near Lake Victoria in Uganda. In Chapter 3, major limitations to phytoplankton primary production in the Fingerponds were high levels of turbidity and total suspended solids. This is likely to affect the productivity of periphyton as well. As a means of determining the viability of this option as a food resource in these ponds three experiments were run to 1) assess the temporal changes in periphyton development on artificial substrates made from bamboo and local wetland plants and determine the underlying factors affecting its development; 2) assess the development of periphyton on substrates not grazed by fish; and 3) determine the impact of artificial substrates on water quality.

Study area and period

This study was carried out in four earthen ponds (each 24 x 8 m, 1.5 average depth) located in Gaba, Kampala (N 0° 14' 59.9", E 32° 38' 14.4") on the northern shores of Lake Victoria, Uganda. The experiments were run in ponds stocked with wild fish; predominantly *Oreochromis niloticus* Linneaus 1758, *O. leucostictus* (Trewavas, 1933) and *O. variabilis* (Boulenger, 1906). Experiment 1 was carried out for a period of six weeks between August and September 2003; Experiment 2 for six weeks in October to November 2003 and Experiment 3, which entailed the use of enclosures, for four weeks between January and February 2004.

Materials and methods

Substrate selection criteria and preparation

Substrates used were: *Phragmites* (*Phragmites mauritianus* (Kunth) reed stems, diameter 0.7-3.5 cm; green bamboo (*Oreobambos* sp.) poles, diameter 1.2-2.5 cm; papyrus (*Cyperus papyrus* L.) culms, diameter 1.3-2.0 cm and *Raphia (Raphia farinifera)* palms. Selection of substrate types was based on their local availability with exception of bamboo which was chosen based on its good performance reported in past studies by Azim *et al.*(2005). The dried substrates were woven into 1 m^2 mats for Experiment 1 and Experiment 2 and 0.5 x 0.5 m mats for Experiment 3. Each mat was hooked on to two strong bamboo (*Bambusa vulgaris*) poles split into two of diameter range 5-8 cm, constituting the plant frame (Figure 4.1).

Sampling protocol

Experiment 1 – Periphyton development in an unmanured pond, preliminary assessment

Thirty-two plant frames with 1 m^2 mats made from *Phragmites*, papyrus, *Raphia* and bamboo were placed in an unmanured pond (pond 3) with reed stems, culms and poles (hereon referred to as strips) lying horizontal, 10 cm below the water surface (Figure 4.1). The pond was stocked predominantly with *Oreochromis niloticus* (2.2 fish m^{-2}). The frames were arranged in 7 rows each consisting of 4 or 5 mats placed from the pond middle to the deep end leaving a free perimeter of 0.5 m, a distance of 0.20-0.25 m between mats and 0.5 m between rows. Each row consisted of at least one mat of each substrate type with frames staggered to allow easy movement of fish.

Each mat was divided into four even sections that were sampled once during the experimental period. Periphyton was sampled weekly starting one week after installation. For periphyton densities, samples were taken from three mats per substrate type selected at random. An area of 5.03 cm^2 (defined by the area of a bottle top) was scraped at 5 and 30 cm from the mat top using a scalpel blade. Samples were pooled depth wise; consequently each sample for analysis was an aggregation of scrapings from three mats. Care was taken not to remove any substrate material. For periphyton species composition, samples from 3 randomly selected mats per substrate type were pooled into a composite sample. The composite sample was transferred into 10 ml vials containing 5 ml of distilled water and preserved immediately with Lugol's iodine solution. All samples were kept in a dark ice cooled box during sampling and transportation to the laboratory.

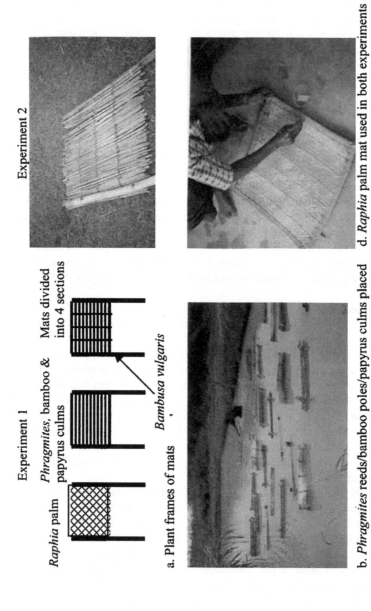

Figure 4.1. Artificial substrate plant frame set up for experiments 1 and 2. Photographs by R.C. Kaggwa.

After sampling, the mats were tagged with coloured string, returned to the pond and excluded from all subsequent sampling. Sub-surface water samples were taken weekly from the pond and from a control (pond 4) from the shallow and middle ends as well as at 10 cm intervals at the deep end. Sampling was done under the cover of a shade. Sampling was done between 10.00 and 13.00 hours.

Experiment 2 (E2) - periphyton development in manured and unmanured ponds
Prior to this experiment, a fish census was carried out using the depletion method (Chapter 2) and fish (*Oreochromis niloticus*, L., *O.variabilis* and *O. leucostictus*) were evenly distributed in mixed sexes and species at a density of 0.5 fish m^{-2}. Fish sizes ranged from 40 to 95 mm and weights from 5 to 30 g. Ten plant frames (1 m^2 each) of *Raphia*, *Phragmites* and bamboo (equal to 60 m^2 of pond surface area i.e. 31 % substrate cover) were suspended vertically (as opposed to Experiment 1 where strips were horizontal) to allow scraping of a whole strip. The mats were put into two ponds manured fortnightly with dry chicken manure at 520 kg ha^{-1} (low manure, LM) and 833 kg ha^{-1} (medium manure, MM). A third pond remained unmanured (NM). The fourth pond acted as a control without substrates and was not manured. The plant frames were arranged in a matrix similar to that of Experiment 1. For ease of sampling and to enable spatial distribution, the rows were demarcated into 3 zones with zone 1 closest to the pond middle and zone 3 closest to the deep end.

Periphyton sampling commenced one week after installation for each substrate type with mats randomly selected from each zone. A strip of known area was scraped from top to bottom using a scalpel blade. Diameters and lengths of reeds/poles for bamboo and *Phragmites* and dimensions for rectangular sections of *Raphia* were determined at each sampling. Samples for ash content and pigment concentrations were taken weekly while samples for species composition determination were taken from the three zones in weeks 1, 3 and 5 and pooled into a composite sample. All samples were treated as in Experiment 1. Sub-surface water samples were taken from each pond fortnightly as described in Experiment 1. Sampling was done between 10.00 and 13.00 hours.

Experiment 3 (E3) – periphyton development on substrates excluded from fish grazing
Four 0.5 x 0.5 m mats made from bamboo, *Phragmites* and *Raphia* were suspended vertically in two cages (1.2x0.6 m) of clear fine nylon mesh which constituted the enclosure. Two enclosures were submerged into the LM, MM and NM pond. Sampling commenced one week after installation for a period of four weeks. Periphyton samples were taken weekly from three mats per substrate type while samples for species composition were taken in weeks 1 and 3. Sampling, preservation and storage were as described for Experiments 1 and 2. Integrated water samples from the upper 30 cm layer were taken weekly from the deep ends of the ponds.

Analytical methods (periphyton, phytoplankton and water quality)
On arrival in the laboratory Chlorophyll *a* (Chl *a*) samples were frozen overnight and analyzed within 24 hours as described by Pechar (1987) (90% acetone: methanol; 5:1 by volume). Phaeophytin *a* was determined using acidification with 0.1 N HCl as described in the same method. Dry matter (105°C, 24 h) and ash free dry matter (AFDM, 550°C, 4-6 h) were determined by weight differences while percentage ash and Autotrophic Index (AI) were computed according to standard methods (APHA, 1995): AI was defined as:

$$\frac{\text{Ash free dry matter } \left(\text{mg m}^{-2}\right)}{\textit{Chlorophyll a } \left(\textit{mg m}^{-2}\right)}$$

(Equation 4.1)

Sedimentation and enumeration of periphytic algae was done by inverted microscope following the modified Utermöl method (Nauwerck, 1963) while other organisms were identified using standard methods (APHA, 1995). Identification was to genus level (Mosille, 1994; John et al., 2002). Cell/organism densities were computed according to Wetzel and Likens (1991).

For phytoplankton, integrated water samples of known volume were collected from the upper 30 cm water layer in the deep ends of the ponds between 10.00 and 14.00 hours and preserved with 1 % Lugol's iodine solution. Sedimentation, enumeration and identification, phytoplankton biomass, Chl a and densities were determined as described in Experiments 1 and 2. Phytoplankton biomass was estimated from bio-volume measurements according to Wetzel and Likens (1991).

In situ measurements for pH, temperature, dissolved oxygen (DO), electrical conductivity (EC) and water transparency measured as secchi depth were taken using WTW handheld meters (MODEL 340i, WTW GmbH, Weilheim, Germany). Underwater light was measured three times in Experiment 1 and twice in Experiment 2, at 10 cm intervals at the pond deep end using a light meter (SKP 200, Skye Instruments Ltd., Llandrindod Wells, UK). Light attenuation was computed based on the equation described in Wetzel and Likens (1991);

$$I_z = I_0 e^{-\eta z}$$

(Equation 4.2)

where I_z = Irradiance at depth z, I_0 = irradiance at surface of pond, η = extinction coefficient / light attenuation (m^{-1}) and z = depth (m).

Water samples filtered through Whatman GF/C filters were used to determine soluble reactive phosphorus (SRP), nitrate (NO_3-N) and phytoplankton chlorophyll a (Chl a). Ammonium nitrogen (NH_4-N) (APHA, 1992), total nitrogen (TN), NO_3-N, SRP and total phosphorus (TP) were determined as described in Chapter 2. All analyses were performed according to standard methods (APHA, 1995).

Statistical analysis
All statistical tests were done at a 5 % probability level using SPSS 11.0 (SPSS Inc, Chicago, Illinois, USA). Due to lack of replicates it was not possible to compare differences between ponds. Repeated measures of Analysis of Variance (ANOVA) was used to test the effect of time (in weeks) and substrate type within ponds for Experiments 1 and 2. Differences between substrate types were tested with the Tukey test. Data sets comprised of periphyton biomass densities, pigment concentrations (Chl a and Phaeophytin a) and ash content. In Experiment 1, samples were taken from two depths (5 and 30 cm) and in Experiment 2 in triplicate per substrate type per pond for six sampling dates.

To determine the key ecological factors affecting pond water quality, factor analysis (Milstein, 1993) was used with factors extracted using the Varimax rotation and principal components calculated from the correlation matrix (Pearson's test). One data set was created with 65 cases from all the experiments. Apart from the water quality and periphyton parameters, some extra variables were created; variables for percentage substrate cover, manure load, pond water level, periphytic algae, zooplankton and other organism densities. Five factors were extracted with the first factor accounting for the highest variation contained in the data set, the second the next highest and so on. The factors are independent of each other, have no unit and are standardized (normal distribution, mean = 0, variance = 1). The coefficients of the linear functions defining the factors were used to interpret their meaning. Coefficients greater than 0.5 were used for interpretation. The sign and relative size of the coefficient indicate the weight placed on each variable.

Results

Visual inspection
In Experiment 1, periphyton colonization was visible on *Phragmites* and bamboo mats within one week of installation while for *Raphia* and papyrus mats it was visible in week 3. At the end of this experiment, bamboo mats were intact while *Phragmites* and *Raphia* mats showed signs of decomposing. Papyrus mats were the least resistant, decomposed rapidly, appeared reddish in colour and gave off a pungent smell by week 6. Mat ridges provided a good surface for trapping sediment and micro fauna more so on *Phragmites* than on the other substrates but mollusca were more common on *Raphia*. In Experiment 2, periphyton colonization was visually evident by the end of week 1 on all substrate types.

Periphyton biomass and productivity
Experiment 1
Periphyton dry matter increased exponentially on substrates within 2 weeks and ranged from 8.1 to 17.7 g m^{-2} for *Phragmites* while for *Raphia* it increased within 3 weeks from 1.2 to 14.0 g m^{-2} (Table 4.1). Meanwhile, dry matter on papyrus decreased from 5.9 to 4.4 g m^{-2} between weeks 1 and 2 but later increased to 11.8 g m^{-2} in week 3. Although the highest dry matter (35 g m^{-2}) was attained on bamboo within 7 days, it declined steadily thereafter. Dry matter and AFDM were higher in the upper mat layer (5 cm) than the lower layer (30 cm) and were higher in Experiment 1 than in Experiment 2 (Figure 4.2 and 4.3).

Ash free dry matter (AFDM) showed similar growth trends with exception of bamboo where an exponential increase occurred within the first three weeks. Mean Autotrophic Index values ± standard error of the mean (SE), over the five weeks were: 881 ± 957 (bamboo), 754 ± 721 (*Phragmites*), 217 ± 187 (*Raphia*) and 216 ± 195 (Papyrus). The effect of week was not significant for any variable while a significant difference between substrates (P<0.05) was noted for dry matter with bamboo attaining the highest values, followed by *Phragmites* (Table 4.1).

Pigment concentrations (Chl *a*) on substrates increased with time to maximum levels in week 4 and dropped to almost the initial values in week 5 for bamboo and *Raphia* while papyrus showed a gradual decline. Chl *a* values ranged from 2.9-80.7 mg m^{-2} (bamboo), 3.3-24.4 mg m^{-2} (papyrus), 1.2-38.7 mg m^{-2} (*Phragmites*) and 2.9-24.4 mg m^{-2} (*Raphia*). The highest values were obtained on bamboo.

Experiment 2
All substrate types demonstrated similar growth of periphyton after one week, but thereafter noticeable differences were observed for bamboo and *Phragmites* in the manured ponds with higher values occurring in the LM pond (Figure 4.3). In this pond, dry matter increased exponentially on all substrate types with bamboo attaining the highest in week 3 (2.0 g m^{-2}) while in the MM pond the highest values were obtained in week 2 (1.5 g m^{-2}). Dry matter on *Phragmites* in the manured

Table 4.1. Periphyton biomass, Experiment 1.

Substrate	DM (g m^{-2})		Chl a (mg m^{-2})		Ash (%)		AFDM (g m^{-2})	
Bamboo	23.88 (13.09)	b	23.87 (4.99)	a	65 (16)	a	5.40 (2.86)	a
Phragmites	18.49 (4.33)	ab	13.53 (6.78)	a	80 (7)	a	3.55 (1.27)	a
Raphia	8.76 (6.16)	a	11.33 (6.55)	a	71 (0)	a	2.09 (1.54)	a
Papyrus	9.47 (3.19)	a	15.16 (7.88)	a	72 (13)	a	2.20 (1.05)	a
Effect of week	ns		ns		ns		ns	
Effect of substrate	*		ns		ns		ns	

Values represent means and in parenthesis standard error, for samples taken in duplicate on five sampling dates for dry matter (DM), pigment concentration (Chl a), percentage ash content and ash free dry matter (AFDM). Samples for week 4 with exception of samples for Chlorophyll a were destroyed accidentally and are not included in the data set. The effect of time defined as weeks and differences between substrates in the same ponds are marked as * for significance level of 5 %. Mean values with the same letter are not significantly different (repeated measures ANOVA).

ponds grew exponentially up to week 4 (3.6 g m^{-2}) but then dropped in week 5 and rose again in the week 6. Temporal variations of dry matter showed exponential increments on bamboo in the 3rd and 4th week for the LM and MM ponds, respectively whilst *Raphia* exhibited the weakest response. Dry matter did not vary significantly between weeks in the ponds with exception of the MM pond. No significant differences were noted between dry matter on different substrates in all ponds (Table 4.2). The lowest dry matter values were obtained in the NM pond.

Pigment concentrations (Chl a and Phaeophytin a) varied between weeks with exception of Chl a in the MM pond. Significant differences ($P<0.05$) were found between substrate types for Chl a and Phaeophytin a in the LM pond only (Table 4.2). In general, Chl a concentrations increased steadily over the first five weeks but dropped by the 6th week. *Raphia* attained the lowest increment of Chl a with values lower than the other substrates. Chlorophyll a and Phaeophytin a were lowest on substrates in the NM pond. Periphyton Chl a ranges in the LM, MM, and NM ponds were: 0.6-18.3 mg m^{-2}, 0.3-15.4 mg m^{-2}, 0.1-2.4 mg m^{-2} (bamboo); 0.3-11.2 mg m^{-2}, 0.2-11.2 mg m^{-2} and 0.1-4.0 mg m^{-2} (*Phragmites*); and 0.2-9.4 mg m^{-2}, 0.3-8.3 mg m^{-2} and 0.1-4.1 mg m^{-2} (*Raphia*).

Ash content and dry matter were not affected by differences in substrate type in all ponds. However, AFDM differed significantly ($P<0.05$) for all substrate types in the NM pond. The Autotrophic Indices (AI) decreased steadily up to week 5 but showed a slight increase for bamboo and *Phragmites* and a sharp rise for *Raphia* in the 6th week. Mean autotrophic index ± standard error for bamboo, *Phragmites* and *Raphia* were: 367 ± 216, 123 ± 40, 615 ± 228 (NM pond); 172 ± 59, 135 ± 32, 859 ± 325 (LM pond); 145 ± 76, 220 ± 93, 435 ± 150 (MM pond).

Periphyton composition

In Experiment 1, ash content on the substrates in the NM pond ranged from 13 to 94 % with the highest value of 94 % obtained on *Raphia* in the second week (Figure 4.2). Ash content on bamboo stabilized after week 2, while for the other substrates it decreased in the next two weeks and rose again in week 5. In Experiment 2, lower values (36-75 %) were obtained on all substrates types. For bamboo and *Phragmites,* the ash content in the manured ponds varied in a similar way whereas for *Raphia* some differences were noted with higher values occurring on mats in the MM pond.

For all substrate types, ash content in the MM decreased up to week 5 followed by an increase in week 6. No significant differences (P<0.05) were found between substrate types in all ponds (Table 4.2).

Periphyton assemblages on all substrates were similar for both experiments and comprised of: algae (Bacillariophyta, Chlorophyta, Cyanobacteria & Euglenophyta); zooplankton (Crustacea, Rotifera & Protozoa) and Fungi in Experiment 1. Less diversity was observed in Experiment 1 (33 genera) compared to Experiment 2 (53 genera). Commonly occurring genera [group (number of genera Experiment 1, number of genera Experiment 2)] included: Bacillariophyta (1, 3) and Chlorophyta (11, 25), Cyanobacteria (6,6), Euglenophyta (1, 3), 1 Protozoa (1,1), Crustacea (7, 8) and Rotifera (6, 7).

Periphytic algal composition varied over time with most groups exhibiting cyclic patterns. Chlorophyta and Cyanobacteria were the dominant algal groups on all substrate types in both experiments. Higher densities of periphyton occurred in Experiment 2 compared to Experiment 1 and showed a 20-fold increment in cell densities. Dominant periphytic algae in Experiment 1 were: *Anacystis* on bamboo (week 1), *Pediastrum* on *Raphia* (week 3), *Scenedesmus* on *Phragmites* (week 1) and *Spirogyra* on papyrus (week 1) (all Chlorophyta); *Anabaena* on bamboo and *Merismopedia* on *Raphia* (week 1) (Cyanobacteria); *Mucor* (a fungus) on bamboo (week 1) and papyrus (week 5). Dominant zooplankton in Experiment 1 included *Moina* on *Raphia* in week 4. In Experiment 2, dominant periphytic algae included: *Ankistrodesmus* (Chlorophyta); *Anabaena*, *Merismopedia* and *Oscillatoria* (Cyanobacteria). Dominant zooplankton in this experiment were: *Rotatoria* (Rotifera) and *Bosmina*, *Cyclops*, *Calanoida*, *Daphnia* and *Diaptomus* (Crustacea).

Euglenophyta were present, although they rarely occurred on *Raphia*. Dominance of periphytic algae and zooplankton periphyton varied for the different substrates with differences between the first and second experiments. In Experiment 1, periphytic algal composition and zooplankton composition were: bamboo and papyrus (48 %, 51 %), *Phragmites* (44 %, 56 %) and *Raphia* (58 %, 42 %); and in Experiment 2 Bamboo (42 %, 58 %), *Phragmites* (40 %, 60 %) and *Raphia* (45 %, 55 %).

Periphytic algal development on the substrates was variable (Figure 4.4). Bamboo showed a general increment in Cyanobacteria densities whilst on *Phragmites* it was high at the beginning but disappeared in week 3 rising thereafter. Chlorophyta density on *Raphia* decreased steadily with time in relation to decreasing secchi depth whereas no distinct trend was observed on papyrus and bamboo. Percentage algal composition for both Cyanobacteria and Chlorophyta were in general lowest in the LM pond and highest in the unmanured pond. *Ankistrodesmus* was dominant on all substrates in all ponds, *Staurastrum* on *Phragmites* (MM pond; week 5), *Gonium* on bamboo, *Chlorella* on *Phragmites* (Week 1; NM pond) and *Selenastrum* and *Stuarastrum* on *Raphia* (week 1 & 5 respectively; NM, pond). *Anabaena* and *Merismopedia* were common to all substrate types in all ponds while other dominant Cyanobacteria were: *Gleotrichia* on bamboo (week 1; LM pond), *Oscillatoria* on bamboo (LM pond), *Phragmites* (MM pond) and *Raphia (NM pond)* all in week 3.

Experiment 1, No manure (Pond 3)

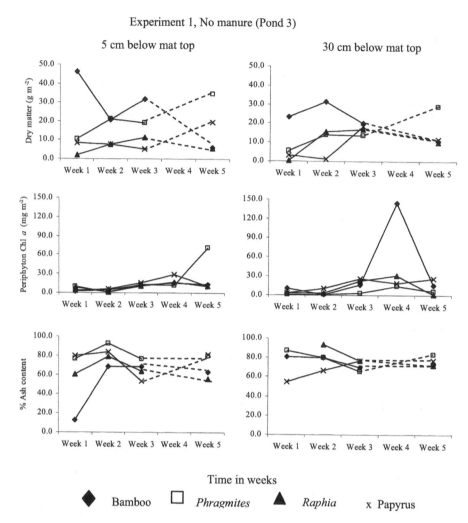

Figure 4.2. Temporal variation of periphyton biomass densities (dry matter), periphyton chlorophyll *a* concentration, percentage ash content, on bamboo, *Phragmites*, *Raphia* and papyrus substrates, Experiment 1, at 5 cm and 30 cm depth below the mat top for five sampling dates. The dotted line between week 3 and 5 indicates that no measurements were taken in week 4 and is only used for completeness of the graph but not for interpretation.

Phytoplankton taxonomic diversity
Phytoplankton genera identified in the two experiments did not vary much and total number of genera found ranged from 24 to 27. For Experiment 1, *Aphanocapsa* and *Merismopedia* (Cyanobacteria) were the most abundant genera while in Experiment 2 it was *Picocyanobacteria* but with *Euglena* having the highest biomass. Some genera types were specific only to the water column and comprised mainly of non-motile species (Table 4.3).

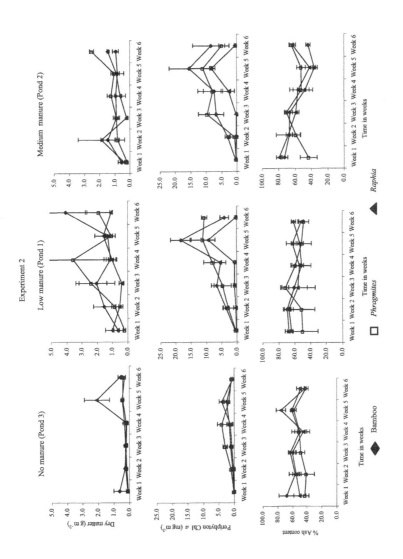

Figure 4.3. Temporal variation of periphyton biomass densities (dry matter), periphyton chlorophyll *a* concentration, percentage ash content, on bamboo, *Phragmites* and *Raphia* substrates, Experiment 2. Sampling done for six sampling dates. Bars indicate standard error of the mean and show variation between samples taken from the same mat.

Table 4.2. Periphyton biomass, Experiment 2

Treatment (Pond no)	Substrate type	Dry matter (g m^{-2})	Chl *a* (mg m^{-2})	% Ash content	Phaeophytin *a* (mg m^{-2})	AFDM (g m^{-2})
NM (Pond 3)	Bamboo	0.36 (0.12) [a]	1.08 (0.34) [a]	56 (6) [a]	0.95 (0.34) [a]	0.15 (0.04) [ab]
	Phragmites	0.27 (0.07) [a]	1.80 (0.46) [a]	49 (6) [a]	1.68 (0.53) [a]	0.12 (0.02) [ab]
	Raphia	0.57 (0.18) [a]	1.08 (0.40) [a]	55 (6) [a]	1.11 (0.43) [a]	0.46 (0.19) [b]
	Effect of week	ns	**	ns	***	**
	Effect of substrates	ns	ns	ns	ns	*
LM (Pond 1)	Bamboo	1.78 (0.70) [a]	6.39 (1.61) [b]	61 (8) [a]	5.46 (1.57) [b]	0.62 (0.19) [a]
	Phragmites	1.67 (0.78) [a]	6.26 (2.10) [b]	55 (11) [a]	5.53 (2.13) [b]	0.61 (0.23) [a]
	Raphia	0.93 (0.29) [a]	1.94 (0.53) [a]	59 (7) [a]	1.25 (0.50) [a]	0.39 (0.14) [a]
	Effect of week	ns	***	ns	***	**
	Effect of substrates	ns	*	ns	**	ns
MM (Pond 2)	Bamboo	0.93 (0.30) [a]	6.85 (3.53) [b]	56 (5) [a]	5.94 (2.97) [a]	0.40 (0.13) [a]
	Phragmites	1.10 (0.15) [a]	5.76 (1.69) [ab]	57 (6) [a]	5.81 (1.78) [a]	0.43 (0.40) [a]
	Raphia	0.91 (0.38) [a]	2.04 (0.43) [a]	61 (8) [a]	1.73 (0.54) [a]	0.26 (0.04) [a]
	Effect of week	*	ns	**	***	***
	Effect of substrates	ns	ns	ns	ns	ns

Values represent means (standard error) for samples taken in triplicate on six sampling dates for dry matter, pigment concentration (chl *a*), percentage ash content, pigment concentration (phaeophytin *a*) and AFDM (ash free dry matter). NM denotes no manure, LM – low manure and MM – medium manure, ns – not significant. The effect of time defined as weeks and differences between substrates in the same ponds are marked as * for significance level of 5 %, ** (1 %) and *** (0.01%). Alpha = 0.05 for all data sets except for Chlorophyll *a*, Pond 2 where 0.10 was used. Mean values with the same alphabetical letter as a superscript are not significantly different.

Table 4.3. Phytoplankton in the water column.

Phyla	Phytoplankton Genera specific to water column		
	All ponds	Experiment 1	Experiment 2
Bacillariophyta			*Cyclotella* (1, 2)
	Aulocoseira	*Nitzschia*	*Nitzschia* (1, 3)
	Cyclostephanos	*Navicula*	*Navicula* (1)
Chlorophyta	*Crucigenia*	*Lagerhemia*	*Epithemia* (1)
	Monoraphidium	*Oocystis*	*Oocystis* (2)
		Tetraedron	*Coelastrum* (1)
		Stuarastrum	*Tetraedron* (1, 2)
			Golenkia (2)
Cyanobacteria	*Aphanocapsa*	*Cylindrospermopsis*	*Cylindrospermopsis* (1, 2)
	Chroococcus		*Anabaenopsis* (2)
	Microcystis		*Coelomoron* (2)
	Picocyanobacteria		*Cyanodictyon* (2)
	Lynbya		*Coelasphaerium* (3)
Euglenophyta	*Trachelomonas*	*Trachelomonas*	*Phacus* (1, 2)
Cryptophyta		*Rhodomonas*	
		Cryptomonas	
Dinophyta		*Peridinium*	

For Experiment 1, samples taken from pond with substrates i.e. pond 3, while for Experiment 2 numbers in parentheses represent pond number.

Water quality

Physical parameters Water quality was stable over the investigative period and showed no spatial differences in the ponds. pH values were near neutral to alkaline with ranges from 7.5 to 8.9; values greater than 7.9 were obtained in the second and third experiments. Electrical conductivity (EC) ranged from 378 to 530 µS cm^{-1} but decreased with time in the manured ponds in Experiment 2. Temperatures ranged from 21.5 to 29.7 °C with higher values occurring in the control. However in the third week of Experiment 1, a sharp difference between the top and bottom temperatures was noted in pond 3 (2.8 C°) and pond 4 (4.5 C°). Alkalinity was high with ranges from 145 to 276 mg CaCO$_3$ L^{-1} in all ponds. Suspended solids (TSS) varied between 32 and 341 mg L^{-1} (Table 4.4) with the unmanured pond having the higher values with resultant low pond transparencies; secchi depths ranged from 8 to 22 cm (Figure 4.5). Vertical profiles for all parameters with exception of TSS and temperature (Experiment 1, week 3), were typical of well-mixed shallow waters.

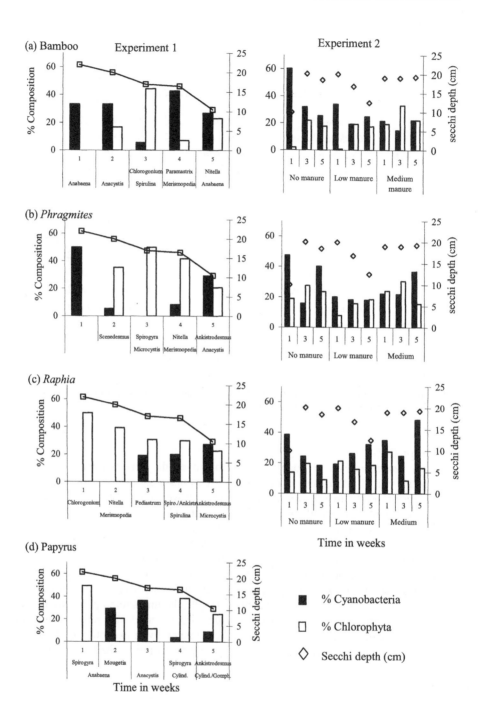

Figure 4.4. Temporal variation of periphytic Cyanobacteria and Chlorophyta composition and secchi depth on substrates. (a) bamboo (b) *Phragmites* (c) *Raphia* and (d) papyrus against time in weeks for Experiment 1 and 2. Values present percentage composition of the algal groups. On the x-axis below time are the dominant Chlorophyta and Cyanobacteria for samples taken in Experiment 1.

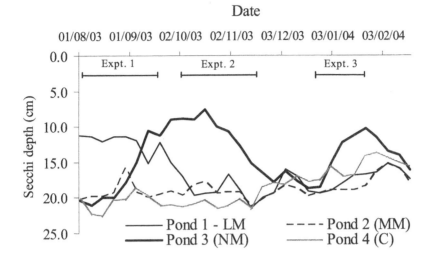

Figure 4.5. Pond transparency daily trends in ponds August 2003-February 2004. LM denotes low manure, MM-medium manure, NM-no manure, C-control. Values plotted are for measurements taken at midday. For Experiments 2 and 3 substrates were put into ponds 1, 2 and 3.

Dissolved oxygen and Chlorophyll a (refer to Table 4.4)
Dissolved oxygen values did not fall below 2 mg L^{-1} throughout the investigative period in all experiments with exception of the LM pond in the last week of Experiment 3, when it dropped to 1.1 mg L^{-1}. Ranges for DO in Experiment 2 were: 4.2 to 9.9 mg L^{-1} (LM) and 2.9 to 10.4 mg L^{-1} (MM), 2.7 to 7.4 mg L^{-1} (NM) and 4.5 to 6.3 mg L^{-1} (control). Phytoplankton Chl *a* concentrations in the ponds were low with values falling below 10 µg L^{-1} and rising to 70 µg L^{-1} in Experiment 1 and Experiment 2 with no noticeable increment in the manured ponds. In Experiment 3, higher Chl *a* values were attained ranging from 10 to 208 µg L^{-1}. In all experiments the control had the lowest Chl *a* concentrations.

Nutrients
Changes in nutrient levels over the investigative period are shown in Figure 4.6 and Table 4.4. Ammonia (NH_4-N) concentrations were higher at the water surface than at the pond bottom but all values were below 0.70 mg L^{-1}. The N:P ratios were below 3.0 in Experiment 1 and ranged from 3.3 to 5.4 in Experiment 2, while in Experiment 3 higher ratios were obtained from 11.5 to 23.5. Dissolved inorganic nutrients were low with NO_3-N ranging from nil to 0.62 mg L^{-1} and SRP <0.100 mg L^{-1}. Higher SRP concentrations were found in NM pond (Experiment 1) and the MM pond (Experiment 2) whereas in Experiment 3, concentrations were within the same ranges in all ponds. In Experiment 1, TN did not vary considerably in all ponds and ranged from 0.25 to 4.74 mg L^{-1}. In Experiments 2 and 3, higher concentrations were obtained; 0.74 to 6.92 mg L^{-1} (Experiment 2) and 0.56 to 9.62 mg L^{-1} (Experiment 3). The MM pond in Experiment 2 had the highest concentrations while in Experiment 3 it was the unmanured pond (NM). TP ranges were: 0.12 to 1.03 mg L^{-1} (Experiment 1), 0.01 to 1.29 mg L^{-1} (Experiment 2) and 0.24 to 1.43 mg L^{-1}

(Experiment 3) with the manured ponds LM and MM attaining higher concentrations.

A positive correlation was found between NH_4-N and periphyton density ($r^2 = 0.30$, P <0.05), while a negative correlation was found between TN and periphyton density ($r^2 = -0.28$, P <0.05) and between SiO_2 and periphytic Chl a ($r^2 = -0.38$, P<0.01). A significant relationship ($R^2 = 0.81$, P<0.01) was found between manure load and periphyton biomass (dry matter) (Experiment 2) for bamboo (Figure 4.7). For the other substrates types, no significant relationships were found. Periphyton Chl a was not significantly affected by manure loads for all periphyton substrate types. However, phytoplankton Chl a increased significantly ($R^2 =0.51$, P<0.05) with increased manure loading showing a linear relationship (Figure 4.7).

Light availability
For Experiment 1, light measurements taken prior to the start of the experiment were high at the beginning but deteriorated with a steeper gradient during the experimental period. Light attenuation values ranged from 11.0 to 20.8 m^{-1} in pond 3 and from 6.6 to 12.3 m^{-1} in the control. In Experiment 2, light availability varied in all ponds with light attenuation values ranging from 9.4 to 15.9 m^{-1} (LM), 5.8 to 14.8 m^{-1} (MM), 7.7 to 14.3 m^{-1} (NM, pond 3) and 3.7 to 7.2 m^{-1} (NM, pond 4). Light intensity varied significantly between sampling dates (Figure 4.8).

Periphyton biomass densities on mats not grazed by fish
Periphyton biomass on mats excluded from grazing by fish ranged from 1-75 g dry matter m^{-2} (bamboo), 1-79 g dry matter m^{-2} (*Phragmites*) and 2-39 g dry matter m^{-2} (*Raphia*). Maximum pigment concentrations (Chl a) were: 2.9 μg cm^{-2} (bamboo), 1.6 μg cm^{-2} (*Phragmites*) and 1.5 μg cm^{-2} (*Raphia*).

Periphyton taxonomic composition on substrates in enclosures, Experiment 3
Differences were noted in the periphytic community for growth of 7 and 21 days with higher genera diversity in the first week. A total of 51 and 36 periphyton genera were identified in weeks 1 and 3, respectively with some similarity but not diversity in groups. Commonly found were: groups (genera number: week 1, week 3); Bacillariophyta (10, 4), Chlorophyta (24, 12), Cyanobacteria (9, 3) and Euglenophyta (3, 2). Other groups found in week 1 were Copepoda (1 genus), Ostracoda (1 genus) and bacteria (3 genera) whilst in week three new groups included Dinophyta (1 genus), Rhodophyta (4 genera), Charophyta (1 genus), Xanophyta (2 genera), Cladocera (3 genera), Rotifera (3 genera) and Fungi (1 genus). *Synedra* (Bacillariophyta) was frequently to dominantly present on bamboo in the NM pond while for the Chlorophyta it was *Volvox* on *Phragmites*. *Spirulina* (Cyanophyta) was rare to abundantly present on all substrate types in both weeks. For other organisms, Nauplii and Ostracods were dominant on bamboo and *Phragmites* (NM pond) while Rotifers were rare to abundantly found on all substrate types for both weeks. Bacteria were abundant on bamboo only in week 1. Periphyton density on ungrazed substrates was higher than for the grazed substrates in Experiment 2 (Table 4.2 and 4.5, Figure 4.9).

Table 4.4. Water quality in ponds for Experiments 1, 2 and 3.

	Experiment 1		Experiment 2				Experiment 3			
	NM + S Pond 3	NM + NS Pond 4	NM + S Pond 3	LM + S Pond 1	MM + S Pond 2	NM + NS Pond 4	NM + NS Pond 4	NM + S Pond 3	LM + S Pond 1	MM + S Pond 2
DO	4.5 ± 0.6	5.5 ± 0.6	5.0 ± 0.6	6.1 ± 0.8	6.0 ± 0.9	5.5 ± 0.2	3.3 ± 0.2	3.4 ± 0.2	3.7 ± 0.4	2.5 ± 0.3
TSS	160 ± 39	69 ± 6	226 ± 21	65 ± 8	96 ± 8	71 ± 8	120 ± 4	262 ± 14	127 ± 19	125 ± 14
Chl a (μg L^{-1})	69.0 ± 23.9	35.9 ± 17.0	45.3 ± 0.7	61.2 ± 9.8	64.0 ± 14.8	20.0 ± 6.8	17.7 ± 3.0	77.0 ± 29.6	123.3 ± 22.7	175.3 ± 18.2
SiO$_2$	12.7 ± 2.2	9.5 ± 0.9	32.4 ± 5.2	13.9 ± 0.8	16.1 ± 1.8	10.6 ± 1.0	25.3 ± 1.9	26.8 ± 2.0	16.1 ± 1.8	12.4 ± 0.6
NH$_4$-N	0.420 ± 0.070	0.380 ± 0.100	0.450 ± 0.050	0.198 ± 0.020	0.290 ± 0.010	0.250 ± 0.058	0.236 ± 0.021	0.286 ± 0.015	0.288 ± 0.036	0.273 ± 0.024
SRP	0.030 ± 0.010	0.023 ± 0.000	0.049 ± 0.000	0.065 ± 0.020	0.122 ± 0.020	0.130 ± 0.006	0.034 ± 0.005	0.027 ± 0.007	0.030 ± 0.007	0.030 ± 0.019
TN	0.64 ± 0.10	1.01 ± 0.27	2.12 ± 0.51	2.66 ± 0.53	3.45 ± 0.89	1.46 ± 0.45	3.02 ± 0.58	4.78 ± 1.96	4.35 ± 0.91	5.10 ± 1.25
TP	0.55 ± 0.06	0.22 ± 0.03	0.64 ± 0.09	0.56 ± 0.10	0.76 ± 0.05	0.27 ± 0.13	0.35 ± 0.06	0.48 ± 0.10	0.62 ± 0.11	0.90 ± 0.25

All units in mg L^{-1} unless otherwise stated. Values represent means ± SE over the sampling period for sub-surface samples taken weekly from the shallow, middle and deep end of the ponds for Experiments 1 and 3 and for weeks 1, 3 and 5 for Experiment 2. LM denotes Low manure, MM-medium manure, NM-no manure and S- Periphyton Substrates. NS denotes no periphyton substrates installed.

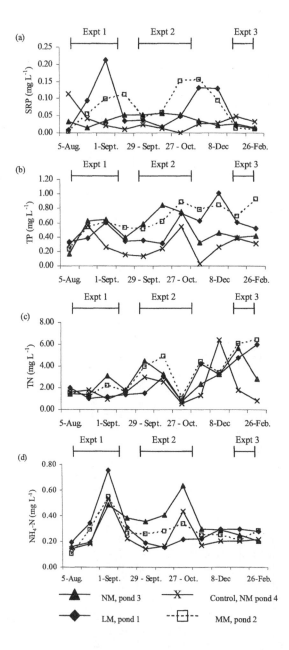

Figure 4.6. Temporal changes of nutrients in ponds over the experimental period (May 2003 to February 2004). Values plotted represent means of sub-surface samples taken from the shallow, middle and deep ends of the ponds during the experimental period. LM denotes Low manure, MM-medium manure and NM-no manure.

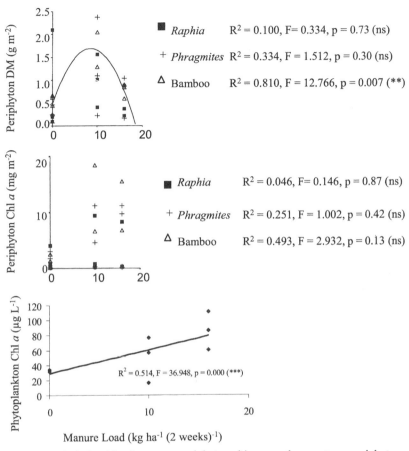

Figure 4.7. Relationship between periphyton biomass dry matter, periphyton chlorophyll a (chl a), phytoplankton Chl a and manure load. Values are for three sampling dates for samples taken fortnightly and in triplicate per substrate type. ns denotes not significant and ** significance level at alpha = 0.1%.

Relationship between pond water quality, periphyton biomass and composition

Five factors contributed over 10% each of the overall variability, together explaining 65.5 % of the variability. The first factor accounted for 18.0 % of the overall variability and comprised of two groups of variables, one with high positive coefficients and the other with negative coefficients. This factor was related positively to phytoplankton Chl a, fish biomass, pH, temperature, TN, manure load, pond water level and periphyton % ash content and negatively to % substrate cover. The second factor (F2, 15.2 %) also had two groups of variables and was related positively to TSS and turbidity and negatively to manure load, pond water level and to some extent secchi depth (all negatively). F3 (11.5%) was related positively to DO, SRP, and periphyton algal and zooplanktonic densities and to some extent to secchi depth while F4 (10.8 %) showed a positive correlation to periphyton biomass (AFDM, dry matter) and the autotrophic index (AI). F5 (10.1%) was positively related to TP, EC and NH_4-N.

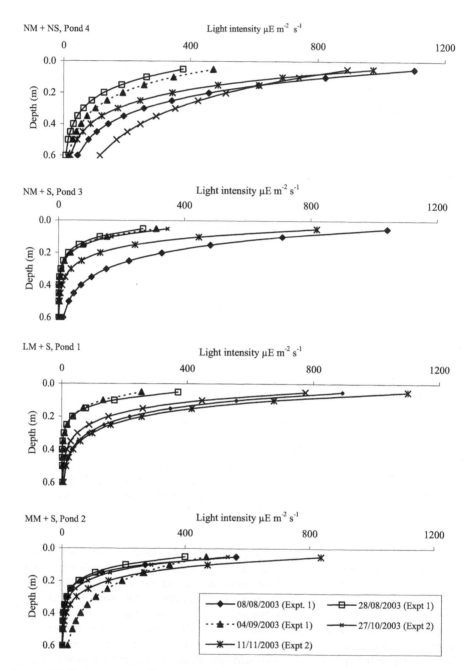

Figure 4.8. Under water Irradiance in ponds, Experiments 1 and 2. NM denotes no manure, NS-no periphyton substrate, S-periphyton substrate, LM- low mnaure and MM- Medium manure

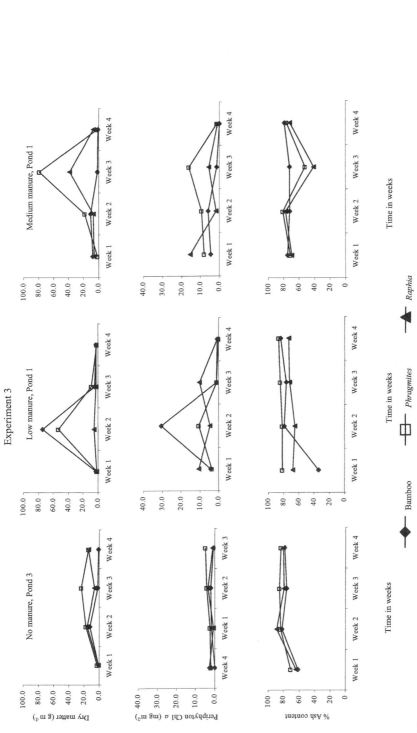

Figure 4.9. Temporal variation of periphyton biomass densities (dry matter), periphyton chlorophyll *a*, percentage ash content on different substrates: bamboo, *Phragmites* and *Raphia*, in enclosures, Experiment 3. Values represent samples taken in triplicate on four sampling dates.

Table 4.5. Mean values (± SE), minimum – maximum for periphyton biomass from enclosures, Experiment 3.

Substrate type	Bamboo			Phragmites			Raphia		
Manure level Variable	No manure	Low manure	Medium manure	No manure	Low manure	Medium manure	No manure	Low manure	Medium manure
Chl a (mg m^{-2})	1.20 (0.44)	9.15 (6.82)	2.98 (1.37)	4.06 (0.95)	3.78 (2.37)	8.44 (3.08)	1.45 (0.39)	5.82 (3.24)	5.77 (2.09)
	0.18 - 2.02	0.8 - 29.43	0.08 - 5.91	2.33 - 6.66	0.06 - 10.49	1.38 - 16.17	0.83 - 2.57	1.54 - 15.31	0.67 - 10.47
Phaeophytin a (mg m^{-2})	0.92 (0.48)	7.12 (5.92)	2.30 (1.17)	3.74 (1.10)	2.94 (1.89)	6.87 (2.72)	1.26 (0.42)	4.31 (1.89)	4.61 (1.48)
	0.13 - 2.30	0.69 - 24.87	0.07 - 5.43	1.37 - 6.64	0.08 - 8.39	1.43 - 14.35	0.57 - 2.45	1.49 - 9.71	0.57 - 7.05
Dry matter (g m^{-2})	3.86 (2.65)	21.02 (17.91)	5.23 (2.48)	14.42 (4.54)	16.30 (12.40)	25.35 (18.43)	9.28 (3.35)	14.59 (8.02)	3.02 (0.77)
	0.56 - 11.74	2.21 - 74.71	0.57 - 10.60	2.33 - 23.85	1.40 - 53.10	1.39 - 79.28	1.50 - 15.81	6.02 - 38.65	2.16 - 5.32
AFDM (g m^{-2})	0.76 (0.45)	4.33 (3.43)	1.33 (0.66)	2.32 (0.61)	2.86 (2.19)	17.00 (15.69)	1.54 (0.47)	9.60 (8.16)	0.92 (0.28)
	0.10 - 2.08	0.37 - 14.59	0.06 - 2.83	0.70 - 3.52	0.21 - 9.36	0.29 - 64.02	0.43 - 2.70	1.24 - 34.09	0.60 - 1.76
% Ash content	75 (4)	68 (12)	75 (2)	81 (3)	83 (1)	70 (6)	79 (4)	65 (8)	70 (2)
	65 - 73	34 - 83	72 - 79	41 - 78	82 - 86	52 - 81	67 - 89	71 - 84	65 - 73

Values represent means of weekly measurements from three substrate types taken in triplicate on four sampling dates for chlorophyll a (chl a), phaeophytin a, dry matter, ash free dry matter (AFDM) and percentage ash content. In parenthesis are the standard error and below the mean value the range minimum-maximum.

Table 4.6. Factor analysis for water quality and periphyton characteristics (N=65)

Variables	Factors				
	F1	F2	F3	F4	F5
Phytoplankton Chl a (μg L^{-1})	**0.785**	0.215	-0.004	0.237	0.142
Fish biomass (kg)	**0.756**	-0.273	-0.209	0.219	-0.126
pH	**0.751**	0.070	0.048	-0.108	0.008
Temperature ($^{\circ}$C)	**0.682**	-0.075	0.070	-0.018	0.002
TN (mg L^{-1})	**0.679**	0.149	-0.060	0.075	-0.273
Manure load (kg ha^{-1} (2weeks)$^{-1}$)	**0.673**	**-0.632**	0.252	-0.007	0.008
% substrate cover	**-0.650**	-0.122	**0.500**	-0.154	0.405
Pond water level (m)	**0.647**	**-0.659**	0.273	-0.001	-0.109
% periphyton ash content	**0.571**	-0.099	-0.020	-0.164	-0.288
TSS (mg L^{-1})	0.088	**0.928**	-0.108	0.035	0.163
Turbidity (NTU)	-0.030	**0.918**	-0.171	0.152	-0.061
Secchi depth (cm)	-0.255	**-0.552**	-0.229	-0.182	**0.513**
DO (mg L^{-1})	-0.022	0.014	**0.897**	-0.021	-0.201
SRP (mg L^{-1})	0.033	-0.220	**0.822**	-0.192	0.238
Zooplankton periphyton (ind. m^2)	-0.156	-0.312	**0.618**	-0.200	0.134
Periphytic algal density (cells m^{-2})	0.167	-0.214	**0.574**	0.070	0.089
Periphyton AFDM (g m^{-2})	0.123	0.051	0.008	**0.934**	-0.012
Periphyton dry matter (g m^{-2})	-0.004	0.094	-0.100	**0.887**	0.038
Autotrophic Index	0.056	0.038	-0.182	**0.793**	-0.238
TP (mg L^{-1})	0.007	-0.213	0.073	-0.235	**0.769**
EC (μS cm^{-1})	-0.133	0.147	0.159	0.142	**0.750**
NH$_4$-N (mg L^{-1})	-0.059	0.490	-0.152	-0.139	**0.713**
SiO$_2$ (mg L^{-1})	-0.144	0.466	0.076	-0.048	-0.077
Periphyton Chl a (mg m^{-2})	-0.102	0.203	0.173	0.148	0.132
Other periphyton densities e.g. bacteria (no. m^{-2})	0.089	-0.035	-0.152	0.042	0.053
% Variance explained	18.0	15.2	11.5	10.8	10.1

F denotes factors and N = number of cases. Values presented in bold have loadings greater than 0.5.

Discussion

Substrate durability and substrate type

The plant types used in this study demonstrated different durability capacities and biodegradability rates. These factors determine how long they can stay in the water without causing adverse effects. Papyrus decomposed quickly probably due to its spongy aerenchyma tissue with abundant air spaces, and caused an increase in suspended solids and a subsequent lowering of pond transparency. Bamboo was the most durable which probably explains its use by other researchers (Hem & Avit, 1994; Ramesh *et al.*, 1999; Keshavanath *et al.*, 2001a).

Periphyton biomass and productivity on grazed substrates

Periphyton development is greatly affected by light availability. As shown in the first experiment, dry matter, AFDM and pigment concentrations were higher in the upper (5 cm) than the lower (30 cm) mat layers. The ponds had a very low transparency (less than 20 cm most of the time) and effective algal growth could only occur near the surface. Deeper mat layers are more suited for attachment of zooplankton and bacteria.

Dry matter production in Experiment 1 peaked on bamboo (35 g m^{-2}) within one week perhaps due to the leaching of nutrients, its good surface for attachment and from fungal colonization that occurred in the first week. Azim *et al.* (2002) reported a first peak within 3 weeks but found the period varied from days to months depending on the ecosystem and substrates. The steady decline in dry matter may be a shift to dominance or loss processes through death, emigration, sloughing and grazing, i.e., the 'loss phase' (Biggs, 1996). With high initial periphyton biomass, the lower layers are out-shaded causing them to deteriorate and become susceptible to sloughing (Muelemans & Roos, 1985). Another reason could be that the fish preferred to graze on periphyton attached to bamboo, leading to a reduction in periphyton biomass with time.

Dry matter values obtained in Experiment 2 were an order of magnitude lower than those in Experiment 1 despite increased nutrient levels in the manured ponds in Experiment 2. From visual inspection, better periphyton development was observed in Experiment 1 as well. The reason for this difference is not clear. It could be due to differences in the sampling methods that caused an over estimation in the first experiment or the high fish grazing pressure in Experiment 2 due to the high recruitment of fish. However, Keshavanath *et al.* (2001b) found no such effects, postulating that when grazed, periphyton is kept in an exponential growth phase. Periphyton productivity of less than 0.66 g AFDM m^{-2} d^{-1} in both our experiments was much lower than the range of 1-20 g AFDM m^{-2} d^{-1} reported by other researchers (Rosemund *et al.*, 2000; van Dam *et al.*, 2002) indicating the presence of limiting factors other than fish grazing.

Assuming 47% C in phytoplankton dry matter (Reynolds, 1984; APHA, 1992; Dempster *et al.*, 1993), the phytoplankton productivity in terms of carbon did not exceed 0.66 g C m^{-2} d^{-1} in any of our ponds which is lower than the 2-4 g C m^{-2} d^{-1} in fish ponds reported by Delincé (1992). Assuming periphyton has a carbon content of 46.5% (APHA, 1995), the maximum periphyton productivity in the ponds was: 0.31 g C m^{-2} d^{-1} (on bamboo). Periphyton doubles the total primary productivity in Experiment 1 and by over 100 % in Experiment 2 in the manured ponds. Contributions of up to 80 % by periphytic algae to lake primary production have been reported by Persson *et al.* (1977) while Azim *et al.* (2002) reported a doubling of primary production in fish ponds.

Periphyton quality and composition on grazed substrates

The ash content of periphyton affects its quality for fish food. The high ash values (up to 80%) in our samples, derived from inorganic material entrapped within the periphyton, however, are unlikely to impair fish growth (Yakupitiyage, 1993). The autotrophic index, AI, with values > 200 implies a majority of non-algal material from heterotrophic associations and non-viable organic material (Huchette *et al.*, 2000; Azim *et al.*, 2002).

The adaptations and mobility of periphytic algae are critical in the development of the periphyton layer on substrates (van Dam *et al.*, 2002; Azim *et al.*, 2005). In this study, periphyton communities evolved with time and showed marked differences in population densities and diversity between the first two experiments. A similar difference between seasons has been reported in other studies (Konan-Brou & Guiral, 1994; Wahab *et al.*, 1999). In Experiment 2, the increment in periphytic algal biomass densities is correlated with enhanced nutrient levels encouraging mono-specific communities that differed between substrate types. This is typical in other fish ponds (Konan-Brou & Guiral, 1994; Huchette *et al.*, 2000) and has been associated with substrate type (Biggs & Smith, 2002) or competition for substrate surface area (van Dam *et al.*, 2002).

The phytoplankton algal composition in the water column comprised mainly of non-motile genera in comparison to the motile periphytic algal genera, e.g. *Microcystis*, *Anacystis* and *Scenedesmus* found on the substrates which arose from the differences in the prevailing environment conditions.

Factors influencing periphyton development

Light, temperature and nutrients

High suspended solids and high turbidities resulted in low pond transparency thus affecting periphyton productivity and structure. Light intensity decreased to almost zero in the upper 40 cm resulting in an increasing dominance of heterotrophic organisms. Periphytic algae such as *Anacystis*, *Microcystis*, *Chlorogonium*, *Spirulina* and *Spirogyra* that dominated the grazed substrates possess qualities that enable them to reposition themselves in the periphyton layer to access more light (Stainer *et al.*, 1971, Bendix, 1960; Halldal, 1962). Temperature variations were small and did not have a major effect on periphyton development.

With manure application, higher periphytic algal densities and higher diversity occurred but the biomass remained low due to the dominance of small-sized genera with low biomass. The low values in the MM pond may also point to self-shading effects. The quadratic relationship between periphyton dry matter and manure load on bamboo suggests that periphyton developed faster and more rapidly on bamboo than on the other substrates.

N:P ratios in the study ponds were low (less than 5) in Experiments 1 and 2 which resulted in dominance of Cyanobacteria (Rhee & Gotham, 1980; Watson *et al.*, 1997; Hargreaves, 1998). Another indicator of N limitation is the dominance of *Anabaena* species that are able to fix nitrogen (Horne, 1977). An interesting observation is the strong correlation between NH_4-N, TN and SiO_2 and periphytic biomass (Chl *a*) which suggests N limitation in both periphyton and pond water. There was no correlation between periphyton biomass and SRP/TP, suggesting that sufficient P was available for the low nitrogen conditions.

Fish grazing

Changes in periphytic algal composition can be indicators of fish grazing. In the enclosures, the higher periphytic algal diversity portrays a distinct colonization pattern over time, i.e. systematic changes in phyla in the absence of fish grazing pressure. The ash content on mats in all the experiments was high and did not vary considerably between the grazed and ungrazed mats indicating similar entrapment of sediment even in the enclosures. The periphyton community was not fully developed

on the substrates in the enclosures, shown by the presence of bacteria and
Bacillariophyta more typical of a young community (Azim *et al.*, 2005). Finally, the
differences in dominant genera on the grazed mats demonstrate a selectivity of fish
for more palatable organisms.

Water quality
Water quality changes in substrate-based ponds were not significant. The highest
variability in water quality arose from the effect of pond inputs/conditions and
enhancement of phytoplankton primary productivity in favourable environmental
conditions (such as temperature and pond water levels). Furthermore, high
suspended solids and turbidity greatly affected the water quality but it reduced
slightly with manure load and higher pond water levels (factor analysis results, F2).
The enhanced periphyton loop in the presence of artificial substrates could be
accompanied by additional surfaces for decomposing bacterial activity thus
improving nutrient (N and P) cycling (Hansson, 1989; Azim *et al.*, 2003). In the
manured ponds, enhanced nutrient availability resulted in higher phytoplankton
which improved sedimentation. Coupled with the substrate surface area which
allowed particle trapping, the pond transparency increased.

DO and SRP play an important role in the synthesis of both algae and
zooplankton on the substrates (F3 (11.5%), Figure 4.10 (c)). Milstein *et al.* (2003)
reported that in the presence of substrates, photosynthesis and heterotrophic activity
by the periphyton contribute to the oxygen balance. In this study these effects are
more evident in the manured ponds which had higher primary productivity. Despite
enhanced nutrient levels in the manured ponds and presence of substrates, the
limitation of light in the ponds could have resulted in periphyton being dominated by
mainly non-algal material.

Nitrification was probably not enhanced by the presence of substrates as no
increments in NO_3-N concentrations were observed. The absence of stimulating
effects of periphyton on nitrification may be due to N limiting conditions which
result in the rapid consumption of any available ammonia and nitrate, thus the
lowering of NO_3-N concentrations in the water column. Furthermore, the substrates
were only present in the ponds for short periods which may not have rendered it
possible for real trends in nitrate accumulation to be observed.

Conclusions

Bamboo, *Phragmites* and *Raphia* can be used to enhance periphyton growth in
Fingerponds and improve availability of natural food. Papyrus, however, is not
suitable as it rapidly decomposes and in these closed systems would result in rapid
deterioration of water quality. Bamboo was superior to the other substrates, followed
by *Phragmites* but since the latter is locally available at almost no cost it is the
preferred option in Lake Victoria wetlands. The 33 % surface area pond coverage by
the substrates did not have an adverse effect on water quality but more research is
required for optimization of periphyton growth. The main limitation to periphyton
productivity in the ponds was the high attenuation of light due to suspended clay
particles in the water despite the application of manure. Preliminary assessment
shows that the fish grazed the periphyton and that fish grazing affected periphyton
composition and biomass but not its productivity. Finally, more knowledge on low
cost measures to reduce turbidity and suspended material in our flood-fed ponds is
required although ponds elsewhere do not necessarily have this problem.

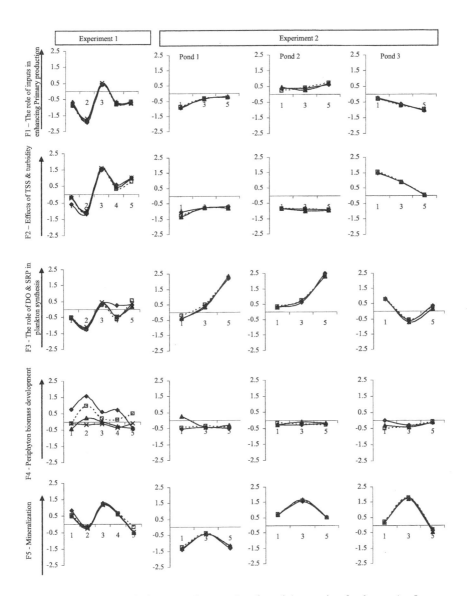

Figure 4.10. Treatment (substrate and manure) and week interaction for factors 1 – 5.

Acknowledgements

This research was funded by the European Union, EU/INCODEV Fingerponds Project No. ICA4–CT–2001–10037 and in part by the Netherlands Government through the Netherlands Fellowship Programme through UNESCO–IHE, Institute for Water Education, Delft, The Netherlands. The research was carried out in partnership with Makerere University Institute of Environment and Natural Resources. We are indebted to the management and staff of National Water and Sewerage Corporation, Uganda for providing technical support and facilities. We are forever thankful for Robert, Ismail, Stephen, Dennis and Gregory who actively participated in field work and laboratory work.

References

APHA, 1992. *Standard methods for examination of water and wastewater.* 18th Edition. American Public Health Association. Washington, DC.

APHA, 1995. *Standard methods for examination of water and wastewater,* 19th Edition. American Public Health Association, Washington, DC.

Azim, M.E. 2001. *The potential of periphyton-based aquaculture production systems.* Ph.D Thesis, Fish Culture and Fisheries Group, Wageningen Institute of Animal Sciences, Wageningen University, The Netherlands. 219 pp.

Azim, M.E., Wahab, M.A., van Dam, A.A., Rooij, J.M., Beveridge, M.C.M., Verdegem, M.C.J. 2002. The effects of artificial substrates on freshwater pond productivity and water quality and the implications for periphyton-based aquaculture. *Aquat. Living. Resour.* 15 (4): 231-241.

Azim, M.E., Vergegem, M.C.J., Singh, M., Van Dam, A.A. and Beveridge, M.C.M. 2003. The effects of periphyton substrate and fish stocking density on water quality, phytoplankton, periphyton and fish growth. *Aquaculture Research* 34: 685-695.

Azim, M.E., Verdegem, M.C.J., van Dam, A.A. and Beveridge, M.C.M. (eds.) 2005. *Periphyton. Ecology, exploitation and management.* CABI Publishing. 319 pp.

Bendix, S.W. 1960. Phototaxis. *Bot. Rev.* 26: 145-208.

Biggs, B.J.F. 1996. Patterns in benthic algae of streams. In: Stevenson, R.J. Bothwell., M.L. and Lowe, R.L. (eds.) *Algal ecology: Freshwater benthic ecosystems.* Academic Press, San Diego, USA. pp. 31-56.

Biggs, B.J.F. and Smith, R.A. 2002. Taxonomic richness of stream benthic algae: Effects of flood disturbances and nutrients. *Limnology and Oceanography* 47: 1175-1186.

Delincé, G. 1992. *The ecology of the fish pond ecosystem with special reference to Africa.* Kluwer Academic Publishers, Dordrecht. 230 pp.

Dempster, P.W., Beveridge, M.C.M. and Baird, D.J. 1993. Herbivory in the tilapia *Oreochromis niloticus*: a comparison of feeding rates on phytoplankton and periphyton. *Journal of Fish Biology* 43: 385-392.

Edwards, P. 1993. Environmental issues in integrated agriculture-aquaculture and wastewater fed fish culture systems: In: Pullin, R.S.V., Rosenthal, H., Maclean, J.H.L. (Eds.), Environment and aquaculture in developing countries. *ICLARM Conf. Prod.,* Vol 31. ICLARM, Manila, pp. 139-170.

Gross, A., Boyd, C.E. and Wood, C.W. 1999. Ammonia volatilization from freshwater fish ponds. *J. Environm. Qual.* 28: 793-797.

Halldal, P. 1962. Taxes. In: R.A. Lewin (Ed.) *Physiology and biochemistry of algae.* Academic Press, New York. pp. 277-296.

Hansson, L.A. 1989. The influence of a periphytic biolayer on phosphorus exchange between substrate and water. *Archiv. Hydrobiol.* 115: 21-26.

Hargreaves, J. A. 1998. Nitrogen biochemistry of aquaculture ponds. *Aquaculture* 166: 181-212.

Hem, S. and Avit, J.L.B. 1994. First results on 'acadja enclos' as an extensive aquaculture system (West Africa). *Bulletin of Marine Science* 55: 1040-1051.

Horne, A.J. 1977. Nitrogen fixation-a review of this phenomenon as a polluting process. *Progs. Wat. Tech.* 8: 359-372.

Huchette, S.M.H., Beveridge, M.C.M., Baird, D.J., Ireland, M., 2000. The impacts of grazing by tilapias (*Oreochromis niloticus* L.) on periphyton communities growing on artificial substrates in cages. *Aquaculture* 186: 45-60.

Jamu, D.M., Chaula, K. and Junga, H. 2003. A preliminary study of the feasibility of using brush parks for fish production in Lake Chilwa, Malawi. *NAGA, World Fish Centre. Quarterly* 26 (1): 4-8.

John, D.M, Whitton, B.A. and Brook, A.J. (eds.). 2002. The Freshwater Algal Flora of the British Isles. *An identification guide to freshwater and terrestrial Algae.* Cambridge University press. 702 pp.

Keshavanath, P., Gangadhar, B., Ramesh, T.J., Beveridge, M.C.M., van Dam, A.A and Verdegem, M.C.J. 2001a. On-farm evaluation of Indian major carp production with sugarcane bagasse as substrate for periphyton. *Asian Fisheries Science* 14: 367- 376.

Keshavanath, P., Gangadhar, B., Ramesh, T.J., van Rooij, J.M., Beveridge, M.C.M., Baird, D.J., Verdegem, M.C.J and van Dam, A.A. 2001b. Use of artificial substrates to enhance production of herbivorous fish in pond culture. *Aquaculture Research* 32: 189-197.

Keshavanath, P., Gangadhar, B., Ramesh, T.J., van Dam, A.A., Beveridge, M.C.M. and Verdegem, M.C.J. 2002. The effect of periphyton and supplemental feeding on the production of indigenous carps *Tor khudree* and *Labeo fimbriatus. Aquaculture* 213: 207-218.

Keshavanath, P., Gangadhar, B., Ramesh, T.J., van Dam, A.A., Beveridge, M.C.M. and Verdegem, M.C.J. 2004. Effects of bamboo substrate and supplemental feeding on growth and production of hybrid red tilapia fingerlings (*Oreochromis mossambicus* x *Oreochromis niloticus*). *Aquaculture* 235: 303-314.

Konan-Brou, A.A. and Guiral, D. 1994. Available algal biomass in tropical brackish water artificial habitats. *Aquaculture* 119: 175-190.

Milstein, A. 1993. Factor and canonical analyses: basic concepts, data requirements and recommended procedures. pp. 24-31 In: M. Prein, G. Hulata & D. Pauly (eds.) *Multivariate methods in aquaculture research: case studies of tilapias in experimental and commercial systems.* ICLARM Stud. Rev. 20. Manila, Philippines. 221 pp.

Milstein, A., Azim, M.E., Wahab, M.A. and Verdegem, M.C.J. 2003. The effects of periphyton, fish and fertilizer dose on biological processes affecting water quality in earthen fish ponds. *Environmental Biology of Fishes* 68: 247-260.

Mosille, O.I.W. 1994. Phytoplankton species of Lake Victoria. In: Reports from the Haplochromis Ecology Survey Team (HEST). Report No. 30B. Zoologisch Laboratorium, Morphology Department, Leiden University, The Netherlands. 84 pp.

Muelemans, J.T. and Roos, P.J. 1985. Structure and architecture of periphytic community on dead reed stems in Lake Maarsseveen. *Archive für Hydrobiologie* 10: 487-502.

Nauwerck, A. 1963. Die Beziehungen zwischen zooplankton und phytoplankton imsec. *Erhen.Symb. Bot. UpSal.* 17 (5). 163 pp.

Norberg, J. 1999. Periphyton fouling as a marginal energy source in tropical tilapia cage farming. *Aquacult. Res.* 30: 427-430.

Pechar, L. 1987. Use of an acetone: Methanol mixture for the determination of extraction and spectrophotometric determination of chlorophyll-*a* in phytoplankton. *Arch. Hydrobiol. Suppl.* 78: 99-117.

Persson, G., Holmgren, S.K., Jansson, M., Lundgren, A., Nyman, B., Solander, D. and Ãnell, C. 1977. Phosphorus and nitrogen and the regulation of lake ecosystems: Experimental approaches in sub-arctic Sweden. In: *Proceedings Circumpolar Conference Northern Ecology,* Ottawa. pp. 1-19.

Ramesh, M.R. Shankar, K.M., Mohan, C.V. and Varghese, T. 1999. Comparison of three plant substrates for enhancing carp growth through bacterial biofilm. *Aquacultural Engineering* 19: 119-131.

Reynolds, C.S. 1984. *The ecology of freshwater phytoplankton.* Cambridge University Press. 384 pp.

Rhee, G. J. and Gotham, I.J. 1980. Optimum N:P ratios and coexistence of planktonic algae. *J. Phycol.* 16: 486-489.

Rosemund, A.D., Mulholland, P.J. and Brawley, S.H. 2000. Seasonally shifting limitation of stream periphyton: Response of algal populations and assemblage biomass and productivity to variation in light, nutrients and herbivores. Canadian *Journal of Fisheries and Aquatic Sciences* 57: 66-75.

Shrestha, M.K. and Knud-Hansen, C.F. 1994. Increasing attached microorganism biomass as a management strategy for Nile Tilapia (*Oreochromis niloticus*) production. *Aquacultural Engineering* 13: 101-108.

Stanier, R.Y., Kunisawa, R., Mandel, M. and Cohen-Bazire, G. 1971. Purification and properties of unicellular blue-green algae (Order Chroococcales). *Bacteriol. Rev.* 35: 171-205.

van Dam, A.A.., Beveridge, M.C.M., Azim, M.E., Verdegem, M.C.J. 2002. The potential of fish production based on periphyton. *Reviews in Fish Biology and Fisheries* 12: 1-31.

Wahab, M.A., Azim, M.E., Ali, M.H., Beveridge, M.C.M and Khan, S. 1999. The potential of periphyton based culture of the native major carp calbaush, *Labeo calbasu* (Hamilton). *Aquaculture Research* 30: 409-419.

Watson, S., McCauley, E. and Downing, J.A. 1997. Patterns in phytoplankton taxonomic composition across temperate lakes of differing nutrient status. *Limnology and Oceanography* 42 (3): 487-495.

Welcomme, R.L. 1972. An evaluation of the acadja method of fishing as practiced in the coastal lagoons of Dahomey (West Africa). *J. Fish. Biol.* 4: 39-55.

Wetzel, R. G. and Likens, G. E. 1991. *Limnological analyses*. 2nd Edition. Springer – Verlag. 391 pp.

Yakupitiyage, A. 1993. Constraints to the use of plant fodder as fish feed in tropical small-scale tilapia culture systems: an overview. In: S.J., Saushik., Luquet, P. (eds.). Fish Nutrition in Practice, 61. *Insitut. National de la Recherche Agronomique*, Les Colloques, Paris France. pp. 681-689.

Chapter 5

Fish production in periphyton based organically manured seasonal wetland fish ponds 'Fingerponds' in Uganda

Abstract

This research investigated the enhancement of fish production through the application of organic manure and presence of artificial substrates in eight 192 m^{-2} seasonal wetland earthen ponds 'Fingerponds' over two grow-out periods between 2003 and 2005. The ponds were predominantly stocked with tilapia (*Oreochromis* spp.) at densities of 0.1 to 0.5 fish m^{-2}. Chicken manure was applied fortnightly at low (521 kg ha^{-1}), medium (833 kg ha^{-1}) and high (1,563 kg ha^{-1}) levels in the first period and at only low levels in the second period. Artificial substrate mats made from *Phragmites*, *Raphia* and bamboo were installed for six weeks in selected ponds in two locations and covered an area equivalent to 31%, 43% or 70% of the pond surface. Periphyton ash content ranged from 21 to 59 %. Periphyton productivities ranged from 0.01 to 0.90 g C m^{-2} d^{-1} resulting in more than doubling of the total primary productivity (periphyton + phytoplankton). Higher net fish yields were obtained in manured ponds in the presence of periphyton substrates and were significantly different ($P<0.05$) between low and medium manured ponds. The highest fish yield of 2,670 kg ha^{-1} was obtained over a 310 day growth period translating into a supply of fish for household consumption of 6.4 kg per person over a 310 day growth period. High recruitment and inbreeding between *O. niloticus* and *O. leucostictus*, dropping water levels, light limitation due to high suspended solids and turbidity and low zooplankton biomass resulted in a lower than possible fish production. Factor analysis indicated that the key environmental factors/processes contributing to greater than 70 % of the overall variance in water quality, pond inputs and fish production were: the role of nitrogen in phytoplankton development, enhancement of primary productivity and sedimentation/re-suspension processes within the ponds. Fingerponds have a high potential for fish production that can be enhanced with improved stock management through reduction of tilapia reproduction.

Key words: Fingerponds, fish yield, manure input, periphyton, tropical semi intensive system, water quality

Publication based on Chapter 5:
Kaggwa, R.C., van Dam, A.A., Balirwa, J.S., Kansiime, F. and Denny, P. 2006. Fish production in organically manured seasonal wetland fish ponds near Lake Victoria, Uganda. Aquaculture Research (submitted).

Introduction

In Africa, chronic hunger continues to be widespread with over 200 million people (28 % of the population) suffering from malnutrition (Clover, 2003). With the rapid increase in population and continuing expectations of growth in the standard of living, pressures on natural resources have become intense. Welcomme (1979) reported declining inland fisheries in this region which have subsequently led to decreased fish consumption. Currently per capita fish consumption rates in Sub-Saharan Africa (SSA) stand at a low 6.7 kg $person^{-1}$ $year^{-1}$ compared to 16 kg $person^{-1}$ $year^{-1}$ for the rest of the world, and are steadily declining (WFC, 2005).

In many societies worldwide, aquaculture has contributed towards alleviating poverty, improving food security and nutritional status of the rural communities particularly where it is traditionally practiced (Edwards, 1999). However, in Sub-Saharan Africa despite numerous resources: abundant natural resources, inexpensive labour, high demand for fish and favourable climate, aquaculture has not been fully exploited. This is mainly due to poor infrastructure and lack of production inputs (Machena & Moehl, 2001). With little traditional aquaculture knowledge there is need to encourage simple technologies that integrate agriculture and aquaculture.

Fingerponds are earthen ponds dug at the edge of natural wetlands and stocked naturally with wild fish during flooding. These systems encourage wise use of wetlands and at the same time provide much-needed protein to resource-poor communities. In order to study their potential for fish production, eight Fingerponds (each 24 x 8 m) were established in two localities along the northern shores of Lake Victoria, Uganda. The ponds were stocked with *Oreochromis* species during the May-June rains in 2003. Preliminary studies showed that the pond environment was suitable for fish survival with dissolved oxygen values maintained above 2 mg L^{-1} with no ammonia toxicity effects. Flood waters used to fill these ponds contained insufficient amounts of essential nutrients and were low in plankton biomass (Kaggwa et al., 2005; Porkornỳ et al., 2005). As a means of enhancing natural food in these systems organic fertilization was applied in combination with artificial substrates for periphyton attachment.

Pond fertilization with animal or green manures has been practiced in other parts of the world to increase fish yields (Green et al., 1989; Egna & Boyd, 1997). This is made possible through three basic pathways: autotrophic, heterotrophic and direct feeding by fish (Colman & Edwards, 1987). Furthermore, the 'periphyton loop' in fish ponds can be exploited through the addition of artificial substrates for periphyton attachment based on traditional fisheries practices such as the 'Acadja' in Benin and the 'Katha' fisheries in Bangladesh (Hem & Avit, 1994; MacGrory & Williams, 1996). Periphyton-based systems have shown promising results in the recent past with net fish yields of up to 2,000 kg ha^{-1} in 135 d^{-1} (Azim et al., 2002).

Organic manure application provides a low cost alternative for enhancing fish production but its indiscriminate use can lead to excessive phytoplankton production which can be detrimental to fish (Pechar, 1995; Dhawan & Kaur, 2002). As primary productivity is enhanced it is important that the food made available is suitable for the fish. *Oreochromis niloticus,* widely distributed in Africa and the world, feeds mainly on algae (including Cyanobacteria) and plant material (Moriarty & Moriarty, 1973; Colman & Edwards, 1987). Cyanobacteria are said to be a better source of nutrition than other algae though some researchers argue that

Cyanobacteria can be toxic (Jos *et al.*, 2005). In view of the foregoing, it was necessary to establish the effects of organic manure application and artificial substrates on the processes driving primary productivity in Fingerponds, as well as determine their impact on fish production. The present study therefore undertook to 1) assess the fish production potential of organically manured ponds in the presence of artificial substrates; and 2) determine the key factors driving fish production in Fingerpond systems.

Study area and period

This study was carried out in eight Fingerponds (24 x 8 m, 1.5 m average depth) located in Gaba (near Kampala, N 0° 14' 59.9", E 32° 38' 14.4") and Walukuba (near Jinja, N 0° 25' 58.1", E 33° 13' 59.8") on the northern shores of Lake Victoria, Uganda. The experiments were run in ponds stocked with *Oreochromis* species at densities of 0.1 to 0.5 fish m^{-2}. The first grow-out period herein referred to as Period 1 lasted from May 2003 to April 2004 while Period 2 lasted from October 2004 to February 2005 in Gaba and from October to April 2005 in Walukuba. Artificial substrates in the form of plant frames were introduced in selected ponds for six weeks between October and November 2003 (Period 1) in Gaba and February to March 2005 (Period 2) in Walukuba.

Materials and Methods

Fish
Fish identification and population census
Fish species were identified on site with the exception of *Haplochromine* species which were preserved in 5 % formalin and identified later in the laboratory. Initial fish population censuses were determined by depletion sampling at the start of each period using a seine net (12 x 2 m, mesh size 6.5 mm). Three successive catches were made in each pond and individual fresh weights and total lengths were determined for fish greater than 5 cm. Fish less than 5 cm were measured in batches of known number. Fish population biomass was estimated by extrapolation of the catches. Classification of fish total length was designated as follows: Class I (0-5 cm), Class II (5.1-10.0 cm), Class III (10.1-15.0 cm), Class IV (15.1-20.0 cm) and Class V (20.1-30.0 cm). Mid-term censuses were carried out in a similar manner while for final harvests, ponds were drained, fishes harvested and weights and total lengths measured. Length-weight regressions of dominant tilapia for the mid-term census populations were calculated using the general model functional linear regression as described by Jensen (1986) as per the log transformed equation:

$$Ln\,W = Ln\,a + b * Ln\,TL \qquad\qquad \text{(Equation 5.1)}$$

where W = mean weight (g), TL- total length (cm), a = intercept and b = slope.

Sex ratio

Sex ratio was determined for fish with total lengths above 10 cm by inspection of the urogenital pores and secondary characteristics, as well as from gonad inspection in preserved material. Gonad states were assessed according to the classification in Ricker (1971). Seven states were applied: state 1 (virgin-immature), state 2 (developing virgin-immature), state 3 (maturing), state 4 (mature or ripe), state 5 (ripe), state 6 (spawning, ripe-running) and stage 7 (spent).

Fish stock management

Following the natural stocking of ponds during seasonal flooding, fish were left to acclimatize to the new environment for a period of six weeks. Thereafter a mid-term census was carried out as described in Chapter 4. To equalize stocks, fish were distributed in equal numbers following the first floods with the number dependent on the overall population catch. In the first grow-out period, Period 1, densities were: 0.4 fish m^{-2} at the Walukuba site and 0.5 fish m^{-2} at the Gaba site with an initial mean weight of 5-30 g and mean total length of 4.0-9.5 cm.

In the second grow-out period, Period 2, no flooding occurred because of the low rainfall and a sudden drop in lake level (Figure 5.1). Here, ponds were stocked by fish that had apparently escaped the final harvest in Period 1. Fish were equalized at densities of 0.1 fish m^{-2} in Gaba and 0.3 fish m^{-2} in Walukuba with mainly males selected through manual sexing in October 2004. To encourage production of zooplankton and zoo-benthic invertebrates, and to reduce the population of small fry, rigorous removal of fish less than 5 cm was carried out monthly using a fine mesh seine net.

Fish gut content and condition factor

To determine the composition of food ingested by the fish, gut content was analyzed during the mid-term census in Period 1 and monthly (starting October 2004) in Period 2. Fish samples were preserved in 10 % formalin for fish with total lengths > 10 cm and 5 % formalin for fish < 10 cm in total length. Gut analysis was carried out in the laboratory using the percentage occurrence method as described by Hyslop (1980) and Balirwa (1998). Fulton's condition factor (K, g cm^{-3}) was calculated for fish at the mid-term census of Period 2 as follows (Ricker, 1975; Wootton, 1990):

$$K = \left(\frac{W}{TL^3} \right) x\ 100 \qquad \text{(Equation 5.2)}$$

where, W is the body weight in g and TL is the total length in cm.

Pond Fertilization and installation of artificial substrates

Dry chicken manure (Nitrogen (N), Phosphorus (P), Potassium (K) values 1.3-1.9%, 1.2-1.4%, 0.1-1.8% in dry matter) purchased from a local market was applied to the ponds fortnightly. The manure was applied into a bamboo crib at one corner of the shallow end at application rates (kg ha^{-1} (2weeks)$^{-1}$) of 521 (low manure, LM), 833 (medium manure, MM) and 1,563 (high manure, HM). During Period 1, different manure levels were employed in the two locations; low manure (LM) and medium manure (MM) in Gaba ponds 1 and 2, respectively while in

Walukuba LM, MM and high manure (HM) were applied to ponds 2, 4 and 3, respectively. Manuring commenced in August 2003 (Gaba) and October 2003 (Walukuba) until March 2004.

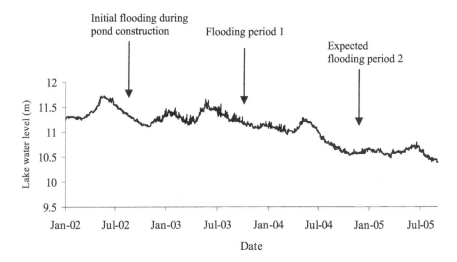

Figure 5.1. Change in Lake Victoria water levels, January 2002-2005 (Source. GOU, 2005a)

Due to low pond water levels in the second period the LM level was applied to three ponds at each location with the fourth pond acting as a control (NM). Manuring commenced in October 2004 in both sites and was continued until February 2005 in Gaba and April 2005 in Walukuba

Artificial substrates made from *Phragmites* (*Phragmites mauritianus* (Kunth) reeds, diameter 0.7-3.5 cm; bamboo (*Oreobambos* sp.) poles diameter 1.2-2.5 cm; and *Raphia (Raphia farinifera)* palm fronds were installed in Gaba (Period 1) in two manured ponds (1 & 2) and one unmanured (pond 3). Substrates were left in the ponds for six weeks between October and November 2003 and covered 31 % of the pond surface area (referred to as Experiment 2 in Chapter 4). At the Walukuba site in Period 2, plant frames from *Phragmites* were introduced into two manured ponds (2 & 4) in February 2005 and covered 43 and 70% of the pond surface area respectively.

Sample collection (periphyton, water and plankton)
All sampling was done between 10.00 and 14.00 hours. Periphyton samples were taken randomly from a known area by the scraping method adopted from Denny and Bailey (1978). Periphyton samples for ash content and pigment concentrations were taken weekly while composite samples for species composition determination were taken fortnightly starting in week 1. Periphyton samples were handled, stored and analyzed as described in Chapter 4. For the determination of periphyton density for each pond in Period 1, the three substrate types were averaged.

Sub-surface water samples were taken monthly from the shallow and middle areas of the ponds and at 10 cm intervals at the deep end. A sample from the whole water column at the deep end was also taken. Sample storage and handling was as described in Chapters 2 and 4.

Phytoplankton samples of known volumes were collected using a 1-litre Van Dorn sampler (Kahl Scienctific Instruments Corp. El Cajon, California, USA) from the upper 30 cm water layer in the deep ends of the ponds and preserved with 1% Lugol's iodine solution. Zooplankton samples were collected from the shallow, middle and deep ends by horizontal hauls using an Apstein net (80 μm mesh) and pooled into a composite sample.

Phytoplankton primary productivity

The mean photosynthetic active radiation (PAR) was derived from incident light radiation recorded every 5 minutes for diel measurements using a light sensor (Data Hog 2, Skye Instruments Ltd, Llandrindod Wells, UK). Underwater light attenuation was measured directly by a light meter (LI-250 A, LI-COR BioSciences, Lincoln, USA). Primary productivity was determined monthly using the modified Winkler's dark and light bottle method (Vollenweider et al., 1974; Wetzel & Likens, 1991) (see chapter 3). Net primary productivity (NPP), gross primary productivity (GPP) and respiration rates were calculated from differences in oxygen concentrations.

Analytical methods

In situ measurements for pH, temperature, dissolved oxygen (DO) and electrical conductivity (EC) were taken using WTW handheld meters (MODEL 340i, WTW GmbH, Weilheim, Germany). Water transparency was measured daily as secchi depth at 12.00 hrs. Total suspended solids (TSS), turbidity, alkalinity, ammonium nitrogen (NH_4-N) (APHA, 1992), total nitrogen (TN), nitrate nitrogen (NO_3-N), soluble reactive phosphorus (SRP) and total phosphorus (TP) were determined according to Standard methods (APHA, 1995).

Periphyton chlorophyll a (Chl a) was determined as described by Pechar (1987) (90% acetone: methanol (5:1, by volume)). Dry matter (105°C, 24h), ash free dry matter (550°C, 4-6h) and the Autotrophic index (AI) were computed according to standard methods (APHA, 1995). AI was defined as:

$$\frac{\text{Ash free dry matter } \left(\text{mg m}^{-2}\right)}{\text{Chlorophyll } a \left(mg\ m^{-2}\right)} \qquad \text{(Equation 5.3)}$$

Sedimentation and enumeration of phytoplankton were done by inverted microscope following the modified Utermöl method (Nauwerck, 1963). Identification was to genus level (Mosille, 1994; John et al., 2002). Zooplankton densities and dry weight biomass were measured according to methods described by Duncan (1975) and Fernando (2002).

Statistical analysis

Statistical analysis was performed using SPSS 11.0 (SPSS Inc, Chicago, Illinois, USA). Differences between plankton biomasses in manured and unmanured systems were tested using the non-parametric Mann Whitney test. Differences between water quality variables were tested using one-way analysis of variance (ANOVA). Where an effect was significant it was followed by a Tukey HSD test.

Differences between means of net fish yields between ponds and location were tested with the t-test.

Direct comparison between treatments was not possible due to lack of replicate ponds. Ecological processes that accounted for the main variability of measured variables were identified through factor analysis (Milstein, 1993). Differences between treatments was shown by the variation in factors over time. Two data sets were created with 68 and 72 cases for the Gaba and Walukuba sites, respectively. Variables used were for water quality parameters as measured. Additional variables were introduced and included: percentage cover of substrates, fish biomass (for which part was extrapolated from actual measurements). An exploratory approach was used for the factor analysis. Four factors were extracted separately for the two locations from the principal components calculated from the correlation matrix using Varimax rotation. Coefficients of the linear functions defining the factors were used to interpret their meaning using sign and relative size of the coefficient as an indication of weight placed upon each variable. Factor loadings with values > 0.500 were considered practically significant and used for interpretation.

Results

Fish stock and species distribution
All ponds were initially self-stocked predominantly with *Oreochromis niloticus* (Linneaus) 1758, *Oreochromis leucostictus* (Trewavas, 1933) and *Oreochromis variabilis* (Boulenger, 1906). Other species found included *Clarias gariepenus* (Burchell, 1822), *Protopterus aethiopicus* (Heckel, 1851), *Gambusia sp.*, *Haplochromines (Astatoreochromis alluaudi* Pellegrin & *Astatotilapia nubila* Boulenger*)* and *Aplochielicthys pumulis* (Boulenger, 1906). *O. leucostictus* was dominant in all ponds during all censuses and final harvests with the exception of the mid-term census in Walukuba where it was *O. niloticus*. Initial stocking densities in Period 1 were 1.1-2.2 fish m^{-2} in Gaba and 0.01-2.1 fish m^{-2} in Walukuba. In Period 2 they were 0.1 fish m^{-2} in Gaba and 0.1-0.9 fish m^{-2} in Walukuba.

Food availability
Plankton biomass in pond water
Zooplankton biomass in all ponds was less than 100 μg L^{-1} and showed little variability between locations. Phytoplankton biomass was variable with distinct differences between ponds and between periods more so in Gaba than Walukuba. The highest phytoplankton biomasses were obtained in Period 2 in the Walukuba ponds (Table 5.1). Differences were noted between phyla; Euglenophyta were significantly higher in manured ponds (Mann–Whitney test, $P<0.05$) in Gaba in Period 1 whilst in Walukuba it was highest in the unmanured ponds in the second period. In the second period in both locations, Cyanobacteria biomasses were highest in the manured ponds. In Walukuba, no significant differences ($P>0.05$) were found between phytoplankton biomasses in the ponds in Period 1. Phytoplankton diversity was similar for both locations and periods; Gaba 26 genera vs. 34 genera (Period 1 and 2, respectively) and 30 genera vs. 35 genera (Period 1 and 2, respectively).

Periphyton quality, composition and biomass

Periphyton assemblages on all substrates comprised mainly of periphytic algae: Bacillariophyta, Chlorophyta, Cyanobacteria and Euglenophyta. Rhodophyta, Xanophyta and Crysophyta were identified in Walukuba in the second period only. Chlorophyta were dominant on substrates in Gaba while in Walukuba it was Cyanobacteria. Higher algal periphytic densities were obtained on substrates in Gaba than in Walukuba (maximum values of 263,000 cells m^{-2} vs. 76,000 cells m^{-2}). In general, development of periphytic algae was dynamic and followed a cyclic pattern with higher periphytic algal diversity occurring on substrates in Walukuba than in Gaba (53 genera vs. 36 genera). Dominant periphytic algae on substrates in Gaba included: A*nkistrodesmus* (Chlorophyta) and *Anabaena*, *Merismopedia* and *Oscillatoria* (Cyanobacteria) while Euglenophyta were rarely present. In Walukuba periphytic algae included: *Anabaena*, *Microcystis*, *Oscillatoria*, *Spirulina*, *Nodularia* (Cyanobacteria), *Fragilaria* (Bacillariophyta) and *Palmella* (Chlorophyta).

Zooplankton abundance was higher in Gaba than in Walukuba (maximum count: 64,000 individuals m^{-2} vs. 1000 individuals m^{-2}). Protozoa were present in the Gaba pond only in the first period (1,000-4,000 organisms m^{-2}). Zooplankton were dominated by Rotatoria (Rotifera) and Bosmina, Cyclops, Calanoida, Daphnia and Diaptomus (all Crustacea) in both locations.

Ash content on the substrates ranged between 21 and 59 % for the two locations. Higher values were recorded in Gaba but they showed no significant differences between ponds for each location (Table 5.2). Autotrophic indices exceeded 1000 on substrates in Walukuba whilst in Gaba they were lower and ranged between 167 ± 70 and 309 ± 117 (mean $\pm$ standard error of mean). Higher pigment concentrations were obtained on substrates in Gaba (maximum mean value of Chl a = 14.6 mg m^{-2} in the LM pond) but dry matter was higher on substrates in Walukuba and ranged between 19.6 and 28.1 g m^{-2}.

Fish diet

The three *Oreochromis* species appeared to have similar diets but some difference was noted in food preference between the two periods (Table 5.3). In Period 2, no insects were found in the fish gut but the guts contained more debris such as sand. Most fish had a high composition of algal detrital material indicating that they fed mainly on the bottom. Cyanobacteria and Chlorophyta were more prevalent in their guts with the exception of the unmanured pond in Walukuba where Bacillariophyta dominated. Fewer fish ingested Cyanobacteria in Gaba Period 2 compared to Period 1 whereas in Walukuba the reverse occurred (Figure 5.2). Only a few guts contained insects (*Chironomidae*) or higher plant material: zooplankton was only found in the first period.

Table 5.1. Plankton biomass (µgL⁻¹) in (a) Gaba and (b) Walukuba, Periods 1 and 2.

Group	Period 1 Manured (n=18) Biomass (µg L⁻¹) Mean	SE	Period 1 Unmanured (n=30) Mean	SE		Period 2 Manured (n=15) Biomass (µg L⁻¹) Mean	SE	Period 2 Unmanured (n=5) Mean	SE	
Gaba										
Cyanobacteria	889	± 155	790	± 183		6,055	± 2,548	480	± 247	*
Cryptophyta	459	± 279	601	± 212		1,431	± 1,056	189	± 103	
Bacillariophyta	469	± 79	5,019	± 4,196		1,097	± 284	104	± 35	*
Dinophyta	319	± 259	169	± 46		350	± 175	0	± 0	
Euglenophyta	6,665	± 2,076	4,272	± 1,525	*	3,463	± 1,942	10,908	± 9,867	
Chlorophyta	191	± 34	8,395	± 8,014		1,776	± 596	205	± 105	*
Zooplankton	110	± 22	55	± 9	**	63	± 17	55	± 22	
Walukuba	Manured (n=21)		Unmanured (n=23)			Manured (n=10)		Unmanured (n=18)		
Cyanobacteria	1,796	± 414	1,628	± 434		44,865	± 18,400	18,582	± 9,576	
Cryptophyta	341	± 117	1,375	± 619		2,687	± 1,525	1,979	± 1,183	
Bacillariophyta	1,691	± 767	1,004	± 474		759	± 370	928	± 284	*
Dinophyta	13	± 11	25	± 13		0	± 0	2,873	± 1,438	**
Euglenophyta	3,252	± 752	2,508	± 671		4,726	± 1,729	22,200	± 11,276	*
Chlorophyta	757	± 112	1,467	± 368		664	± 166	1,850	± 474	*
Zooplankton	55	± 11	65	± 9		17	± 2	26	± 9	

Values represent means for each period. In Gaba Period 1, samples taken for twelve months: two ponds manured for nine months; two ponds unmanured for nine months; and four ponds unmanured for three months. In Period 2, samples taken for four months: three manured ponds and one unmanured pond. In Walukuba Period 1, samples taken for eleven months; three manured ponds and one pondunmanured for seven months and four ponds unmanured for four months. In Period 2 samples taken for seven months: all ponds unmanured only in the first month, thereafter only one unmanured. Significance differences denoted by * for P value < 0.05 and ** P value < 0.01.

Table 5.2. Periphyton density on artificial substrates in Walukuba and Gaba Fingerpond sites.

Substrates	Gaba (Period 1)									Walukuba (Period 2)					
	Combination of Bamboo, *Phragmites* & *Raphia*									*Phragmites only*					
Pond no	Pond 3 (NM)			Pond 1 (LM)			Pond 2 (MM)			Pond 2 (LM)			Pond 4 (LM)		
Statistic	Mean	SE	CV	Mean	SE	CV	Mean	SE	CV	Mean	SE	CV	Mean	SE	CV
% substrate cover	31			31			31			43			70		
Chl *a* (mg m^{-2})	4.0	1.3	78	14.6	5.3	89	14.6	4.9	83	7.6	2.1	94	8.5	6.0	165
Phaeophytin (mg m^{-2})	3.7	1.3	8	12.2	4.5	89	13.5	5.1	93	7.2	2.1	95	8.0	6.4	183
Ash %	40.0	3.0	5	59.0	1.7	2	58.0	3.7	5	21.1	5.8	78	21.1	6.2	64
DM (g m^{-2})	1.2	0.4	76	4.4	0.8	43	2.9	0.6	47	28.1	5.8	54	19.6	8.4	97
AFDM (g m^{-2})	0.7	0.3	100	1.6	0.3	52	1.1	0.2	53	22.0	2.3	58	11.4	1.6	66
AI	309	117	93	208	83	97	167	70	102	**			**		

Values represent means for samples taken weekly over a six weeks for each period, standard error of mean (SE) and coefficient of variance (C.V.). LM denotes low manure and MM - medium manure. ** represents Autotrophic Index (AI) with values greater than 1000. Three substrate types installed in each pond in Period 1; bamboo, *Phragmites* and *Raphia* while in Period 2, substrates made from *Phragmites* were selected as the most suited for Fingerponds.

Table 5.3. Fish gut content in Gaba and Walukuba sites, periods 1 and 2.

Location	Gaba		Walukuba	
	1 (2003/04)	2 (2004/05)	1 (2003/04)	2 (2004/05)
Period (Year)				
Number of fish sampled	56	21	39	61
Stomach fullness	E, Q, H, F	E, H, F	E, Q, H, F	E, Q, H, F
Gonad state	3 to 6	1 to 6	2 to 6	2 to 6
% Females	50	33	40	31
% Males	50	67	60	69
FW (g): Min-Max (Mean)	8-132 (41)	22 -129 (67)	12-49 (25)	9-79 (30)
% No. of *Oreochromis niloticus*	55	19	59	12
% No. of *O. leucostictus*	45	10	41	36
% No. of *O. variabalis*	0	72	0	52
Total length (cm): Min-Max (Mean)	10.0-25.5 (11.3)	10.5-20.5 (14.8)	9.0- 22 (11.3)	7.5-15.9 (11)
Standard length (cm): Min-Max (Mean)	0.8-15.5 (9.6)	7.5-14.5 (11.1)	2.0-11.0 (8.6)	5.5-14.5 (8.6)
Food composition	% Number of fish containing each item (% food composition)			
Higher plant material	5.4 (15 %)	14.3 (2-50 %)	35.9 (5-17 %)	8.2 (2-50%)
Detritus & sedimented algae	33.9 (19-100%)	19 (48-100%)	35.9 (12-100%)	55.7(0-100 %)
Debris e.g. sand	1.8 (<5 %)	76.2 (2-30 %)	5.1(2 – 4 %)	28.00%
Fry				1.6 (100%)
Chironomids	5.4 (<5 %)		7.7 (9 %)	

Samples were taken at the mid-term census for Period 1 and monthly in Period 2. Stomach fullness is classified as E-empty, Q-quarter, H-half, F-full. FW represents fresh weight (g), min-minimum and max-maximum.

Fish growth and yield

Following the mid-term censuses, stock evening and commencement of manuring, fish increased in number and size in both locations but with distinct differences in the two periods. In the first grow-out period (Period 1) in Gaba, fish increased from a mean weight of 6.8 to 55.5 g (manured ponds) and from 4.9 to 22.0 g (unmanured ponds) over a culture period of 220 days. In Period 2, mean weights rose from 58.0 to 81.7 g (manured ponds) and from 58.1 to 101.8 g (unmanured ponds) over 123 days. In Walukuba, Period 1, mean fish weights rose from 20.0 to 46.8 g (manured ponds) and from 20.0 to 37.2 g (unmanured ponds) over 185 days. In Period 2, weights increased from 26.5 to 36.6 g (manured ponds) but decreased in the unmanured ponds from 42.7 to 35.1 g over 179 days.

Final harvests were dominated by small fish (total length, TL < 5 cm) mainly *O. leucostictus* in both periods and locations with proportions for Period 1 and Period 2 ranging from 12 to 42 % and from 2 to 69 % in Gaba and from 1 to 10 % and 10 to 44 % in Walukuba. *Oreochromis* spp. accounted for 39-100 % of the overall fish biomass in the two periods. In Walukuba they accounted for 79-100 %. Excessive recruitment occurred in all ponds with fish fry and fingerlings (TL < 5 cm) accounting for a significant percentage of the total fish biomass; 26%, 40 % and 20% (ponds 1, 2 & 3 respectively, Period 1) and 80 and 61% (ponds 1 & 3, Period 2) in Gaba with none found in the other ponds. In Walukuba they accounted for 89%, 26%, 82% and 30% (Period 1) and 62%, 35%, 78% and 32% (Period 2) for ponds 1, 2, 3 and 4 respectively.

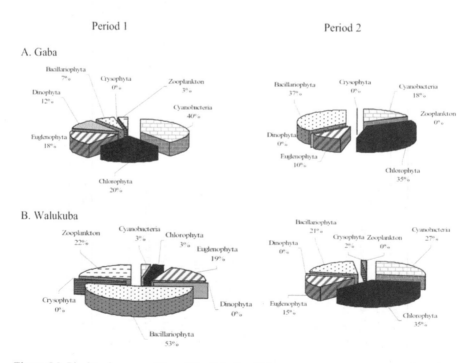

Figure 5.2. Planktonic composition of the fish diet. Values represent percentage number of fish found with specific plankton groups in their guts as indicated. Number of samples for Period 1 and 2 were 56 and 21 in Gaba and 59 and 61 in Walukuba, respectively.

Male to female ratios in the mid-term census and final harvest census for the first period were 1:1 and 1:1.3 (Gaba), 1.2:1 and 1:1.6 (Walukuba) respectively. In the second period they were 1:2 and 7:1 (Gaba) and 4:1 and 2:1 (Walukuba). Condition factors (Ks) calculated for fish in the mid term censuses ranged from 1.5 to 3.5 g cm^{-3} (n = 87, Gaba) and 1.1 to 3.6 g cm^{-3} (n = 308, Walukuba).

Five total length (TL) classes of fish were identified ranging from 0.1 to 30.1 cm for both locations and both periods (Figures 5.3a and b). In the final harvests, more fish with TLs > 10 cm were found in the ponds in Period 1 compared to Period 2 and occurred mostly in manured ponds. Length-weight regressions for *Oreochromis* spp. for the mid-term censuses showed little differences between locations, periods and fish types (Table 5.4).

Fish yields obtained from the two locations varied considerably with differences between location, pond inputs and periods (Table 5.5). Clearly higher net fish yields (NFY) were achieved in manured ponds in both locations with significant differences (P<0.05) between the two locations. Differences between periods were significant for Gaba but not for Walukuba. The highest yield of 2,670 kg ha^{-1} was attained in the pond that received medium manure in Gaba in the first grow-out period (a total growth period of 310 days, with manure applied after the first 110 days). The lowest yield of 30 kg ha^{-1} was obtained in the second grow-out period of 186 days in Gaba in a pond with low manure. High manure levels applied to pond 3 in Walukuba resulted in a low fish yield (Figure 5.4a). Higher NFYs were obtained in ponds that had periphyton substrates in place for part of the culture season. Higher substrate densities resulted in higher fish yields (Figure 5.4b).

Table 5.4. Length - weight relationships of fish at mid-term censuses, periods 1 and 2.

Location Fish species	Period 1 (Mid-term census) Length-weight relationship variables						Period 2 (Mid-term census) Length weight relationship variables					
	Mean L (mm)	Mean W (g)	a	b	r^2	n	Mean L (mm)	Mean W (g)	a	b	r^2	n
(a) Gaba												
O. niloticus	66.5	7.0	0.03	2.98	0.95	83	176.8	118.6	0.10	2.46	0.98	32
O. variabilis	86.6	13.0	0.08	2.34	0.83	12	162.2	79.4	0.05	2.65	0.89	30
O. leucostictus	67.2	7.6	0.03	2.84	0.99	164	126.2	55.5	0.05	2.65	0.98	25
(b) Walukuba												
O. niloticus	121.7	37.1	0.03	2.92	0.98	75	135.9	52.9	0.12	2.33	0.92	95
O. variabilis	101.7	18.5	0.01	3.09	0.85	48	133.3	46.3	0.09	2.42	0.91	159
O. leucostictus	85.6	13.4	0.01	3.42	0.99	182	110.4	29.4	0.05	2.65	0.83	54

Note: L represents fish total length (mm), W-fish mean weight (g), r-correlation coefficient, n-number of individual fish sampled, a-regression intercept and b-regression slope.

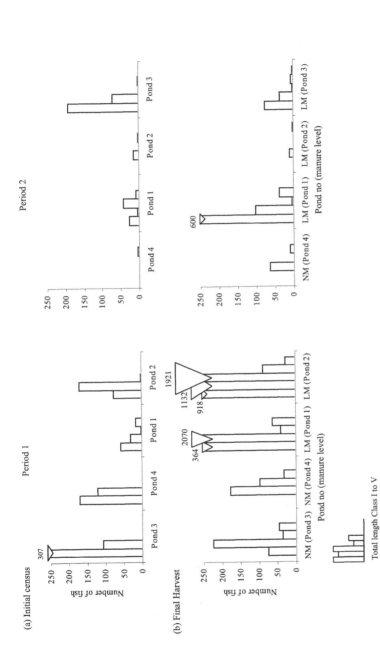

Figure 5.3a. Length-weight classification for fish in Gaba, Periods 1 and 2. (a) initial stock and (b) final harvest. Bar charts represent number of fish in ponds as per selected Class range. Class I refers to fish with total length (cm) ranging from 0.0 to 5.0 cm, Class II (5.1-10.0 cm), Class III (10.1-15.0 cm), Class IV (15.1-20.0 cm) and Class V (20.1-30.0 cm). Bar labels with a triangular symbol indicate the number of fish for bars that exceed the Y-scale.

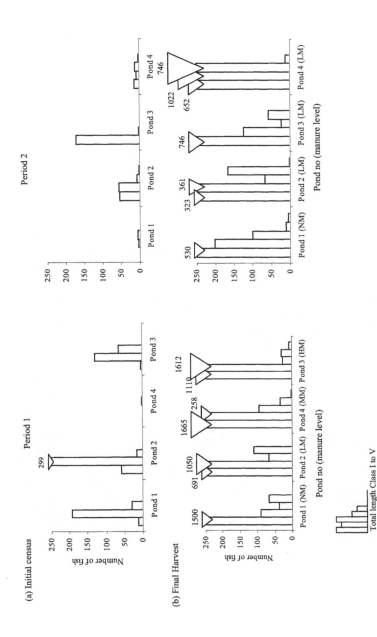

Figure 5.3b. Length-weight classification for fish in Walukuba, Periods 1 and 2. (a) initial stock and (b) final harvest. Bar charts represent number of fish found in ponds as per selected Class range. Class I refers to fish with total length (cm) ranging from 0.0 to 5.0 cm, Class II (5.1-10.0 cm), Class III (10.1-15.0 cm), Class IV (15.1-20.0 cm) and Class V (20.1-30.0 cm). Bar labels with a triangular symbol indicate the number of fish for bars that exceed the Y-scale.

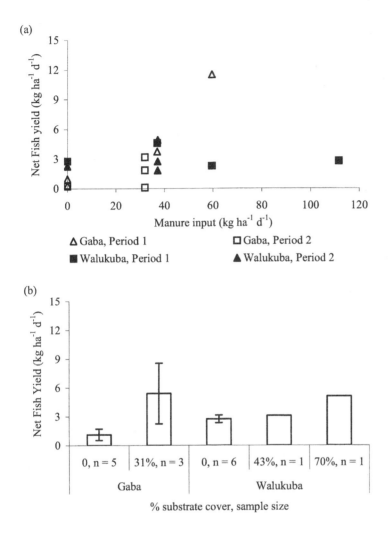

Figure 5.4. Relationship between net fish yield (kg ha^{-1} d^{-1}) and (a) manure input (kg ha^{-1} d^{-1}) (b) percentage substrate cover. Values plotted for b) are means ± standard errors of the mean for ponds with 0, 31, 43 and 70 % substrate cover over the substrate investigative period.

Table 5.5. Fish population biomass and yields for Gaba and Walukuba, Periods 1 and 2.

Period (GPBR, GPR, d)	Period 1 (90, 220 d)		Period 2 (63, 123 d)					
Treatment (Pond no.)	NM* (3)	4 (NM)	1 (LM*)	2 (MM*)	4 (NM)	1(LM)	2 (LM)	3 (LM)
Initial population	419	266	205	278	11	19	14	11
Mid season population	369	328	900	1009	5	128	3	834
Stock equalization population	100	100	100	100	20	20	20	20
Initial Biomass (g)	1,012	781	3,580	1,714	196	382	332	526
Mid season Biomass (g)	930	589	1,502	2,155	407	2,633	246	1,926
Stock equalization Biomass (g)	604	604	604	604	907	813	547	606
Stocking densities (fish m^{-2})	2.18	1.39	1.07	1.45	0.06	0.10	0.07	0.06
Redistributed densities (fish m^{-2})	0.52	0.52	0.52	0.52	0.10	0.10	0.10	0.10
Periodically harvested biomass (kg ha^{-1})	-	-	-	-	8	91	12	125
Final Biomass (kg ha^{-1})	252	108	891	2,670	69	344	30	133
Gross fish yield (kg ha^{-1} d^{-1})	1.10	0.47	3.89	11.66	0.63	3.54	0.34	2.10
Net fish yield (kg ha^{-1} d^{-1})	0.96	0.34	3.76	11.52	0.24	3.19	0.11	1.84
Manure input (kg ha^{-1} d^{-1})	0	0	37	60	0	32	32	32

(b) Walukuba

Period (GPBR/GPR, d)	Period 1(34, 185 d)				Period 2 (105, 179 d)			
Treatment (Pond no.)	1 (NM)	2 (LM)	4 (MM)	3 (HM)	1 (NM)	2 (LM*)	3 (LM)	4 (LM*)
Initial population	250	410	2	222	10	119	173	34
Mid season population	267	295	2	150	1443	537	246	2503
Stock equalization population	70	70	70	70	60	60	60	60
Initial Biomass (g)	2,410	2,455	456	2,147	242	925	284	481
Mid season Biomass (g)	1,921	2,847	515	2,541	3,969	3,672	1,742	5,732
Stock equalization Biomass (g)	1,300	1,300	1,300	1,300	1,224	1,794	1,426	967
Stocking densities (fish m^{-2})	1.30	2.14	0.01	1.16	0.05	0.62	0.90	0.18
Redistributed densities (fish m^{-2})	0.36	0.36	0.36	0.36	0.31	0.31	0.31	0.31
Periodically harvested biomass (kg ha^{-1})	-	-	-	-	259	246	282	366
Final Biomass (kg ha^{-1})	675	1,079	677	568	344	30	133	69
Gross fish yield (kg ha^{-1} d^{-1})	3.08	4.93	3.09	2.59	2.47	3.09	2.09	5.10
Net fish yield (kg ha^{-1} d^{-1})	2.77	4.62	2.78	2.28	2.25	2.76	1.83	4.92
Manure input (kg ha^{-1} d^{-1})	0	37	112	60	0	37	37	37

GPBR represents growth period before redistribution, GPR-growth period after redistribution, LM-low manure, MM-medium manure, *presence of substrates (six weeks period). Yields (kg ha^{-1} d^{-1}) are for GPR. Pond 4 Walukuba, Period 1 was not naturally stocked but had two fish introduced into the pond following stocking.

Water quality

Water quality in the ponds over the two periods was generally stable within a period though some differences were found between periods (Table 5.6a and b). Total suspended solids (TSS) were high in all ponds and resulted in low pond transparencies (secchi depth less than 20 cm) (Figure 5.5). pH ranged from neutral to alkaline (7.0 to 9.7). Temperatures ranged from 22.0 to 29.1°C. Pond water alkalinity in both locations was high (174 to 561 mg $CaCO_3$ L^{-1}) with higher values occurring in Walukuba. Pond water levels were higher in Gaba than Walukuba; maximum monthly means were 0.9 m (Gaba) cf. to 0.6 m (Walukuba). However, in both locations lower water levels were observed in Period 2. EC, TSS, NPP and GPP differed significantly (P<0.05) between ponds in both periods. Phosphorus (TP and SRP), DO, Chl a, and zooplankton biomass differed significantly (P<0.05) between the ponds in Gaba in Period 2. In Walukuba differences between ponds were significant for TSS and SRP in the first period.

(a) Gaba

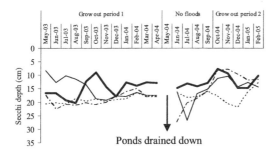

Ponds drained down

(b) Walukuba

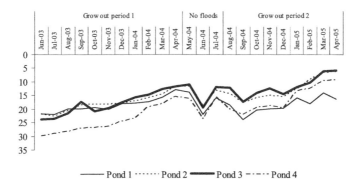

—— Pond 1 ······ Pond 2 ▬▬ Pond 3 –·–·· Pond 4

Figure 5.5. Changes in pond transparency (secchi depth, cm) over the study period. Values represent monthly means for daily measurements.

Table 5.6a. Water quality, Gaba, Periods 1 and 2

Variable	Period 1 (n = 12)			
	NM (Pond 3)	NM (Pond 4)	LM (Pond 1)	MM (Pond 2)
pH	8.4 ± 0.10 a	8.3 ± 0.08 a	8.0 ± 0.12 a	8.4 ± 0.11 a
EC (µS cm⁻¹)	437 ± 7.60 b	458 ± 12.52 b	346 ± 13.78 a	458 ± 8.57 b ***
DO	4.6 ± 0.23 a	4.9 ± 0.21 a	4.1 ± 0.44 a	4.6 ± 0.46 a
Alkalinity (mg CaCO₃ L⁻¹)	222 ± 2.22 b	227 ± 1.70 b	174 ± 2.18 a	224 ± 1.70 b
TSS	187 ± 18.45 b	113 ± 8.17 a	146 ± 20.95 ab	122 ± 19.16 a *
Turbidity (NTU)	237 ± 25.00 b	163 ± 23.00 ab	187 ± 35.00 ab	122 ± 18.00 a
NO₃ - N	0.10 ± 0.01 a	0.07 ± 0.01 a	0.07 ± 0.01 a	0.15 ± 0.02 a
NH₄ - N	0.39 ± 0.02 a	0.31 ± 0.01 a	0.36 ± 0.02 a	0.34 ± 0.01 a
TN	3.21 ± 0.17 a	2.41 ± 0.12 a	3.26 ± 0.10 a	3.83 ± 0.12 a
TP	0.46 ± 0.07 ab	0.32 ± 0.08 a	0.57 ± 0.06 ab	0.63 ± 0.08 b *
SRP	0.036 ± 0.01 a	0.037 ± 0.01 a	0.057 ± 0.01 a	0.054 ± 0.01 a
Chl a (µg L⁻¹)	41.1 ± 6.07 a	38.9 ± 14.37 a	46.9 ± 8.73 a	83.3 ± 19.02 a
NPP (g O₂ m⁻² d⁻¹)	4.2 ± 1.48 a	2.7 ± 0.70 a	4.9 ± 1.16 a	11.7 ± 2.49 b *
GPP (g O₂ m⁻² d⁻¹)	6.3 ± 1.57 a	3.2 ± 0.89 a	6.8 ± 1.35 a	13.3 ± 2.29 b *
Zooplankton biomass (µg L⁻¹)	57 ± 17.25 a	53 ± 12.07 a	114 ± 29.84 a	74 ± 17.68 a

Variable	Period 2 (n = 5)			
	NM (Pond 4)	Pond 1 (LM)	Pond 2 (LM)	Pond 3 (LM)
pH	8.6 ± 0.03 a	8.2 ± 0.22 a	8.9 ± 0.10 a	8.8 ± 0.20 a
EC (µS cm⁻¹)	747 ± 54.91 b	395 ± 13.30 a	835 ± 19.24 b	521 ± 31.97 a ***
DO	5.0 ± 0.13 a	5.1 ± 0.63 a	7.2 ± 0.15 b	7.1 ± 0.68 b **
Alkalinity (mg CaCO₃ L⁻¹)	312 ± 3.74 b	193 ± 3.05 a	322 ± 2.96 b	218 ± 2.90 a ***
TSS	332 ± 24.18 b	248 ± 40.56 ab	117 ± 14.05 a	218 ± 64.78 ab ***
Turbidity (NTU)	442 ± 44.00 b	176 ± 55.00 ab	278 ± 29.00 a	110 ± 38.00 a
NO₃ - N	0.03 ± 0.00 a	0.01 ± 0.00 a	0.01 ± 0.00 a	0.01 ± 0.00 a
NH₄ - N	3.46 ± 0.26 a	2.55 ± 0.33 a	2.04 ± 0.14 a	1.68 ± 0.13 a
TN	10.29 ± 0.45 a	14.89 ± 0.66 a	12.22 ± 0.05 a	11.81 ± 0.43 a
TP	0.33 ± 0.12 a	1.45 ± 0.23 b	1.58 ± 0.23 b	1.82 ± 0.29 b
SRP	0.07 ± 0.02 ab	0.034 ± 0.01 a	0.14 ± 0.04 b	0.035 ± 0.02 a *
Chl a (µg L⁻¹)	51.7 ± 9.68 a	98.9 ± 27.20 ab	157.0 ± 25.08 b	171.6 ± 24.76 b **
NPP (g O₂ m⁻² d⁻¹)	0.2 ± 0.07 a	2.2 ± 1.00 a	10.5 ± 2.24 b	11.1 ± 1.52 b ***
GPP (g O₂ m⁻² d⁻¹)	0.4 ± 0.06 a	5.3 ± 1.27 a	13.3 ± 2.78 b	15.5 ± 1.50 b ***
Zooplankton biomass (µg L⁻¹)	24 ± 8.75 a	59 ± 26.31 ab	119 ± 30.03 b	37 ± 15.33 ab *

Units are in mg L⁻¹ unless otherwise stated. Values are means ± standard errors of means for samples taken monthly over the grow-out period. Statistical comparisons using ANOVA were made separately for each location between values of the same row and period. * <0.05, ** < 0.01, ***<0.001. Values with the same superscript letter are not significantly different.

Four factors were extracted and accounted for 52.7% (Gaba) and 50.9% (Walukuba) of the overall variance in the water quality, pond inputs and fish production data (Tables 5.7 and 5.8). Similarities were found between the locations for three of the four factors. The first factor (F1) in Walukuba and F4 in Gaba (17% and 11 % of the overall variability, respectively) both showed strong positive correlation between alkalinity, EC and pH and negative with SiO_2. F2 (13 %) in Gaba and F4 (9%) in Walukuba showed strong positive correlation with NPP and GPP. For Gaba, this factor showed a strong positive correlation with fish biomass and manure load as well. Both F3 (13%) were strongly correlated (positively) with TSS and turbidity and negatively with secchi depth. In Gaba, F3 was correlated with NH_4-N whilst in Walukuba it co-existed with water level. F1 (16%) in Gaba was strongly positively correlated with nutrients (TN and TP), algal biomass (Chl a) and density, temperature and DO and negatively with water level. F2 (13%) in Walukuba was positively correlated with nutrients (TN, SRP and NO_3-N) and percentage substrate cover and negatively correlated with zooplankton biomass.

Table 5.6b. Water quality, Walukuba, Periods 1 and 2

Variable	NM (Pond 1)	LM (Pond 2)	MM (Pond 4)	HM (Pond 3)
	Period 1 (n = 11)			
pH	8.7 ± 0.07 [a]	9.0 ± 0.05 [a]	9.1 ± 0.07 [a]	9.1 ± 0.09 [a]
EC (μS cm^{-1})	688 ± 33.27 [a]	834 ± 38.85 [a]	1007 ± 21.51 [a]	965 ± 30.94 [a]
DO	5.5 ± 0.79 [a]	5.4 ± 0.76 [a]	5.1 ± 0.55 [a]	5.9 ± 0.86 [a]
Alkalinity (mg CaCO$_3$ L^{-1})	510 ± 4.67 [a]	626 ± 7.30 [a]	662 ± 3.31 [a]	683 ± 7.30 [a]
TSS	136 ± 13.81 [b]	141 ± 13.69 [b]	88 ± 7.67 [a]	180 ± 25.39 [b] ***
Turbidity (NTU)	83 ± 10.00 [a]	95 ± 10.00 [a]	99 ± 16.00 [a]	67 ± 12.00 [a]
NO$_3$ - N	0.01 ± 0.00 [a]	0.00 ± 0.00 [a]	0.01 ± 0.00 [a]	0.00 ± 0.00 [a]
NH$_4$ - N	0.31 ± 0.02 [a]	0.31 ± 0.01 [a]	0.29 ± 0.01 [a]	0.34 ± 0.01 [a]
TN	3.56 ± 0.13 [b]	3.53 ± 0.22 [a]	3.33 ± 0.12 [a]	5.06 ± 0.22 [a]
TP	0.60 ± 0.04 [a]	0.80 ± 0.12 [ab]	0.59 ± 0.07 [a]	1.13 ± 0.32 [b] **
SRP	0.050 ± 0.02 [a]	0.100 ± 0.06 [ab]	0.090 ± 0.03 [b]	0.150 ± 0.08 [b]
Chl a (μg L^{-1})	96.9 ± 24.02 [a]	89.9 ± 23.92 [a]	58.5 ± 15.27 [a]	121.9 ± 82.02 [a]
NPP (g O$_2$ m^{-2} d^{-1})	5.4 ± 1.63 [a]	8.3 ± 1.47 [a]	5.7 ± 1.50 [a]	5.4 ± 1.33 [a]
GPP (g O$_2$ m^{-2} d^{-1})	7.5 ± 1.96 [a]	9.6 ± 1.03 [a]	6.4 ± 1.62 [a]	7.8 ± 1.67 [a]
Zooplankton biomass (μg L^{-1})	82 ± 9.82 [a]	82 ± 15.31 [a]	39 ± 3.93 [a]	69 ± 13.36 [a]
Variable	NM (Pond 1)	LM (Pond 2)	LM (Pond 3)	LM (Pond 4)
	Period 2 (n = 7)			
pH	8.9 ± 0.10 [a]	9.1 ± 0.06 [a]	9.2 ± 0.09 [a]	9.4 ± 0.12 [a]
EC (μS cm^{-1})	786 ± 39.83 [a]	1209 ± 105.84 [a]	1315 ± 115.57 [a]	1287 ± 67.18 [a]
DO	5.3 ± 0.23 [a]	6.5 ± 0.30 [ab]	7.0 ± 0.65 [ab]	7.0 ± 0.85 [b] *
Alkalinity (mg CaCO$_3$ L^{-1})	503 ± 15.51 [a]	732 ± 12.81 [a]	801 ± 12.68 [a]	720 ± 5.99 [a]
TSS	155 ± 12.82 [a]	227 ± 71.17 [a]	201 ± 65.57 [a]	223 ± 29.85 [a]
Turbidity (NTU)	68 ± 8.00 [a]	176 ± 40.00 [a]	91 ± 31.00 [a]	89 ± 19.00 [a]
NO$_3$ - N	0.01 ± 0.00 [a]	0.02 ± 0.00 [a]	0.02 ± 0.00 [a]	0.02 ± 0.00 [a]
NH$_4$ - N	0.77 ± 0.04 [a]	1.04 ± 0.02 [a]	0.91 ± 0.04 [a]	1.06 ± 0.02 [a]
TN	27.62 ± 1.98 [a]	35.66 ± 1.90 [a]	20.71 ± 0.60 [a]	35.05 ± 1.90 [a]
TP	0.92 ± 0.14 [a]	3.76 ± 0.76 [b]	4.08 ± 0.68 [b]	3.44 ± 0.83 [ab] **
SRP	0.030 ± 0.01 [a]	0.860 ± 0.41 [ab]	0.630 ± 0.39 [ab]	1.200 ± 0.59 [b]
Chl a (μg L^{-1})	98.7 ± 15.31 [a]	292 ± 38.18 [ab]	297 ± 92.84 [b]	266 ± 73.20 [ab] *
NPP (g O$_2$ m^{-2} d^{-1})	3.9 ± 1.06 [a]	6.3 ± 1.15 [a]	7.2 ± 1.22 [a]	2.5 ± 1.99 [a] *
GPP (g O$_2$ m^{-2} d^{-1})	7.3 ± 0.94 [a]	7.8 ± 0.83 [a]	10.5 ± 1.44 [a]	5.8 ± 1.31 [a]
Zooplankton biomass (μg L^{-1})	32 ± 2.32 [ab]	64 ± 9.14 [a]	35 ± 7.74 [ab]	17 ± 1.98 [a]

Units are in mg L^{-1} unless otherwise stated. Values are means $\pm$ standard errors of means for samples taken monthly over the grow-out period. Statistical comparisons using ANOVA were made separately for each location between values of the same row and period. * <0.05, ** < 0.01, ***<0.001. Values with the same superscript letter are not significantly different.

Table 5.7. Factor analysis results (water quality, pond inputs, fish production) for Gaba, Period 1 and 2

Variables	F1	F2	F3	F4
Water level (m)	**-0.731**	0.028	-0.461	-0.196
TP (mg L^{-1})	**0.726**	0.242	0.260	0.077
TN (mg L^{-1})	**0.716**	0.052	0.163	0.168
Chl a (μg L^{-1})	**0.695**	0.351	0.110	0.153
DO (mg L^{-1})	**0.658**	-0.218	-0.052	0.214
Algal density (count L^{-1})	**0.639**	0.066	-0.219	0.037
Temperature (^{o}C)	**0.620**	0.151	-0.400	0.217
Fish biomass (kg ha^{-1})	-0.164	**0.884**	-0.004	-0.050
NPP (g O$_2$ m^{-2} d^{-1})	0.211	**0.863**	-0.169	0.152
GPP (g O$_2$ m^{-2} d^{-1})	0.313	**0.823**	-0.143	0.115
Manure load (kg ha^{-1} (2weeks)$^{-1}$)	0.079	**0.803**	-0.235	-0.140
TSS (mg L^{-1})	0.073	-0.096	**0.895**	0.120
Secchi depth (cm)	-0.181	0.104	**-0.846**	0.122
Turbidity (NTU)	-0.129	-0.266	**0.806**	-0.011
NH$_4$-N (mg L^{-1})	0.047	-0.087	**0.572**	0.563
Alkalinity (mg CaCO$_3$ L^{-1})	0.095	0.017	0.000	**0.894**
EC (μS cm^{-1})	0.336	-0.029	0.058	**0.853**
SiO$_2$ (mg L^{-1})	-0.131	-0.046	0.243	**-0.563**
pH	0.466	0.241	0.104	**0.554**
N:P	0.003	-0.128	0.183	0.246
SiO$_2$:P	-0.328	-0.275	-0.038	-0.072
Algal biomass (μg L^{-1})	-0.014	-0.105	0.065	0.076
Percentage substrate cover	-0.003	0.047	-0.113	-0.170
Zooplankton biomass (μg L^{-1})	-0.028	0.127	-0.163	0.070
SRP (mg L^{-1})	0.209	-0.021	-0.030	0.167
NO$_3$-N (mg L^{-1})	-0.240	-0.034	-0.045	-0.285
% Variance Explained (N = 68)	15.7	13.0	12.6	11.4

F denotes factors, N = number of cases. Loadings in bolded are greater than 0.500

Table 5.8. Factor analysis results (water quality, pond inputs, fish production) for Walukuba, Period 1 and 2.

Variables	F1	F2	F3	F4
SiO$_2$:P	**-0.889**	-0.034	-0.183	-0.027
Alkalinity (mg CaCO$_3$ L^{-1})	**0.757**	0.129	0.351	-0.066
EC (μS cm^{-1})	**0.745**	0.205	0.286	0.091
SiO$_2$ (mg L^{-1})	**-0.718**	0.062	0.142	-0.278
pH	**0.620**	0.280	0.108	0.003
TP (mg L^{-1})	**0.584**	0.499	0.175	-0.259
Manure load (kg ha^{-1} (2weeks)$^{-1}$)	**0.568**	0.103	0.038	0.209
N:P	**-0.566**	0.379	-0.008	0.219
TN (mg L^{-1})	-0.005	**0.765**	0.376	0.081
Zooplankton biomass (μg L^{-1})	0.039	**-0.671**	0.186	-0.134
Percentage substrate cover	0.188	**0.651**	0.208	-0.148
SRP (mg L^{-1})	0.449	**0.589**	0.238	-0.119
NO$_3$-N (mg L^{-1})	0.115	**0.547**	0.391	-0.089
Turbidity (NTU)	0.063	0.132	**0.852**	-0.141
TSS (mg L^{-1})	0.227	0.101	**0.809**	-0.130
Water level (m)	-0.290	-0.054	**-0.694**	0.232
Secchi depth (cm)	-0.187	-0.334	**-0.627**	0.039
NPP (g O$_2$ m^{-2} d^{-1})	0.031	-0.135	-0.164	**0.844**
GPP (g O$_2$ m^{-2} d^{-1})	0.063	0.099	-0.296	**0.766**
Temperature (^{o}C)	0.090	0.087	0.012	**0.618**
Algal density (count L^{-1})	0.003	0.226	0.116	0.101
Algal biomass (μg L^{-1})	0.152	-0.065	0.029	-0.250
NH$_4$-N (mg L^{-1})	0.220	0.483	0.039	0.042
Fish biomass (kg ha^{-1})	0.041	-0.006	0.240	0.157
DO (mg L^{-1})	0.021	0.096	0.015	-0.084
Chl a (μg L^{-1})	0.389	0.388	0.256	-0.219
% Variance Explained (N = 72)	17.2	12.6	12.5	8.6

F denotes factors, N = number of cases. Loadings in bolded are greater than 0.500

Discussion

Fish densities, condition and recruitment

Stocking densities have a major effect on fish yields and stunting of fish (Delincé, 1992; Glasser & Oswald, 2001). In the first period, ponds were naturally stocked with a mixed sex polyculture with stocking densities of 1 to 2 fish m^{-2}. In Fingerponds in Kenya which were totally submerged during flooding, more fish entered the ponds and stocking densities were as high as 12.6 fish m^{-2} (Porkorný et al., 2005). In Gaba and Walukuba, ponds were not inundated (Chapter 2). Densities of 0.1 to 1 tilapia m^{-2} have been shown to be effective in attaining fish of good sizes at harvest as long as there is a constant supply of food (Glasser & Oswald, 2001).

At the mid-term censuses, condition factors (Ks) for the fish were greater than 1 (1.1- 3.6 g cm^{-3}) which tallies with findings for Nile tilapia (de Graaf, 2004) and indicates that the fish had adequate food. Manuring of ponds resulted in improved condition factors. A positive relationship has been reported between condition and spawning (Jalabert & Zohar, 1982; Brummett, 1995; Lowe-McConnell, 1982) whilst excessive recruitment and/or breeding lead to low fish production (McGinty, 1985; Lovshin et al., 1990). In our ponds, many small fish of less than 25 g and total lengths (TLs) of 8-10 cm, particularly, O. leucostictus, were sexually mature and precocious spawning was observed.

During Period 2, manual sexing of fish, and stocking only male fish, plus monthly removal of female fish and small fish (< 5 cm) were done to curb reproduction. These measures however, were not very effective as errors are easily made when sexing tilapia <30 g (Delincé, 1992) but the ratio of male to female fish was increased. Periodic harvesting removed fish fry but not the eggs; overcrowding created competition for food, slowed down growth rates and resulted in stunted fish with fewer 'table' sized ones (total length about 15 cm). In the littoral zone of Lake Victoria stunted fish have been observed (Lowe-McConnell, 1979; Balirwa, 1998). Theoretically, fish in our ponds could have been stunted right from the start.

Food availability

Availability of natural foods in ponds is crucial for fish production (Teichert-Coddington & Green, 1993; Knud-Hansen & Lin, 1993). Manuring of the ponds resulted in increased nutrient levels and enhanced primary productivity. Phytoplankton primary productivity was limited by high suspended solids and turbidity in the ponds (Chapter 3). Furthermore, phytoplankton composition in the ponds was dictated by low zooplankton densities, poor light conditions (due to high TSS and turbidity), low N:P ratios and concentrated nutrient levels (due to low water levels). These factors provided a conducive environment for Cyanobacteria, which can assimilate dissolved atmospheric nitrogen and migrate vertically (Bíró, 1995). It is possible that some Cyanobacteria species were not palatable to the fish although tilapias have been observed to digest them (Moriarty & Moriarty, 1973).

The artificial substrates provided additional food and fish were observed to graze on the periphyton. With periphyton ash ranges of 21-59 %, a reasonable nutritional value for fish was provided. An ash content of less than 30 % is reasonable in nutritional terms (Yakupitiyage, 1993). Periphyton productivity in Gaba and

Walukuba ranged from 0.01 to 0.09 g m^{-2} d^{-1} and from 0.02 to 4.15 g m^{-2} d^{-1} respectively. Assuming 47% C in phytoplankton dry matter (APHA, 1992) the periphyton productivity did not exceed 0.04 g C m^{-2} d^{-1} (Gaba) and 0.90 g C m^{-2} d^{-1} (Walukuba). Periphyhton contributed from 9 to 65 % of the total primary productivity (periphyton + phytoplankton) in Gaba and 2 to 92 % in Walukuba. The total primary productivity increment in our manured ponds more than doubled.

The fish in the ponds fed mainly on detrital material irrespective of body size. Juvenile fish mainly feed on zooplankton (Robotham, 1990; Porkorný et al., 2005) and later switch to phytoplankton (Moriarity & Moriarity, 1973; Trewavas, 1983; Elhigzi et al., 1995). The combination of low zooplankton biomass (less than 100 µg L^{-1}) and the high recruitment levels resulted in a low transfer of energy from phytoplankton to zooplankton. Furthermore, high densities of Cyanobacteria did not help boost zooplankton biomass either as they are of poor food value to zooplankton (Bíró, 1995).

Environmental factors and pond inputs, and their effect on fish production

Water temperatures of 22.0 to 29.1°C and pH ranges of 7.9-9.7 are normal for tropical fish ponds (Wurts & Durborow, 1992; Boyd, 1997; Hargreaves & Heusel, 2000). They did not show much variation between periods. High turbidity, notably in Gaba, limited light penetration and photosynthesis leading to reduced phytoplankton primary productivity. In Gaba, this turbidity arose mainly from inorganic clay particles while in Walukuba, it was mainly due to phytoplankton (Chapter 3). Despite low phytoplankton primary productivity, dissolved oxygen levels in the ponds were maintained at acceptable levels; above 2 mg L^{-1} at 06.00 hours and above 4 mg L^{-1} between 10.00 and 14.00 hours. However, the high Cyanobacteria densities in our ponds could lead to deterioration of water quality (e.g. low oxygen levels due to sudden algal crashes) which could trigger fish kills.

Unionized ammonia (NH_3) can reach toxic levels for fish in high pH waters (Boyd, 1990). In Fingerponds, NH_3 is likely to predominate in the late afternoon and early evening when pH values are highest. Since this period is short, no toxicity was experienced during the investigative period. However, extra caution must be taken when thick algal blooms develop as this causes a rise in pH and subsequent release of NH_3.

The key processes explaining the variability in water quality and fish production, in relation to pond inputs and conditions were similar for ponds in both locations. In Gaba, 16 % of the variability was explained by the synthesis of algal biomass and pond water levels Factor 1 (F1). As the pond water level decreased, particularly in the second period, higher phytoplankton Chl a concentrations were obtained. Enhanced phytoplankton primary productivity due to manure application resulted in increased fish biomasses (F2). In Walukuba, fish biomass was not related to either manure input or primary productivity. Evidently, manure input was more important for fish production in Gaba than in Walukuba. The epilithic periphyton on the rock bottoms in Walukuba could have caused a higher natural primary productivity in these ponds hence the reduced impact of manure. On the other hand, the significantly higher fish yield obtained in the medium manured pond in Gaba in the first period, may over-emphasize the effect of manure on phytoplankton primary productivity.

One would expect the highest fish yields to come from the highly manured pond in Walukuba. However, the yield from this pond was relatively low pointing to likely limitation of fish growth. Without replicates it can only be speculated that the low fish yields in this pond arose from high algal self-shading effect which lowered net phytoplankton primary productivity and hence fish yields.

Sedimentation/re-suspension processes (F3) played a special role in our ponds. In Walukuba, this factor was affected by water level while in Gaba it was not. This reinforces the results of different sources of turbidity in the two locations. In Gaba, fine clay particles remain suspended in the water column irrespective of water levels whilst in Walukuba, the organic turbidity due to phytoplankton is mostly re-suspended from the pond bottom when water levels are low.

In Period 2 in Walukuba, nutrient levels were increased in the presence of substrates but zooplankton biomass in the water column decreased. The negative effect of enhanced nutrients on zooplankton biomass may be explained by the higher biomass of large algae (mostly colonial or filamentous), which leads to interspecific competition between different zooplankton types and size-selective predation. This phenomenon has been observed in eutrophic lakes (Wetzel, 1983).

Fish growth performance and yields

During the first period, it was apparent that manure input had a direct effect on fish biomass in Gaba whereas in Walukuba this was unclear. The factor analysis confirmed this (cf. F2, Gaba and F4, Walukuba). In the second period, growth was greatly slowed in both locations with a decrease in mean weight of fish in the unmanured ponds in Walukuba. The slow growth is likely to be due to stunting of the fish or the high feeding pressure from the hundreds of small fish. Large numbers of female fish tend to use up their energy in production of gametes which could explain stunting. Certainly, food limitation, high fish densities towards the end of the culture period and variable environmental conditions are among factors playing a role in reduced growth (Noakes & Balon, 1982; Lorenzen, 2000).

The *Oreochromis* species which were dominant in the ponds share a similar diet. Furthermore, *O. niloticus* is able to interbreed with *O. leucostictus* resulting in a hybrid slightly bigger than *O. leucosticus* but still smaller than *O. niloticus* (GOU, 2005b). The high competition for food and space by the dominant fish species hampered their growth. Despite the low average fish size (55 g), some fish greater than this appeared in the final harvests. The normal size range at harvest for *Oreochromis spp.* from semi-intensive ponds in Uganda after an 8 months grow-out period is 100-200g (GOU, 2005b). With high inbreeding in our ponds, it was not possible to get a large proportion of bigger tilapia, i.e. 150-450 g fish[-1] but total fish yields are comparable with other studies in Africa (Delincé, 1992; Egna & Boyd, 1997).

Assuming that one pond belongs to one household of 8 persons, the maximum net fish yields (NFYs) obtained in the two locations can be translated into 6.4 kg person[-1] period[-1] in Gaba and 2.6 kg fish person[-1] period[-1] in Walukuba. With decreasing per capita fish consumption rates in the region, Fingerponds can contribute significantly to household protein supply.

Management options

Following natural stocking of ponds, initial censuses should be carried out within a month of the pond disconnecting from the lake water. This will enable easy determination of the actual stocking densities prior to the precocious spawning which is bound to happen. Manual sexing of fish should be done for fish sizes greater than 30 g and all males returned to the ponds. All fish with total lengths less than 5 cm should be removed periodically. This can be done every three to four weeks and will provide protein for the family as well. Bearing in mind that Fingerponds are low-cost systems that are manageable by resource-poor farm households, the options for curbing recruitment in these natural systems are limited. Effective predation e.g. by *Clarias gariepinus* may be the solution to curbing precocious spawning. It also provides extra protein since it is eaten by the local people. Where *Clarias* does not stock the pond naturally, it can be obtained from the wetland-lake edge. More research is required on the effectiveness of *Clarias spp.* as a predator although other studies show an effective ratio of *Clarias* to tilapias of 1:2.7 in combination with the African snakehead (*Ophiocephalus obscuris*) (de Graaf, 1996). As pond water levels decrease over the grow-out period, manure inputs need to be regulated and reduced accordingly. It is recommended that the current manure regime of every application fortnightly be maintained with the addition of green manures (e.g. vegetables). Substrates for periphyton growth provide a good source of additional food for the fish and may be installed in the ponds. The surface area used in this study of 30-70 % of pond surface area can be maintained.

Conclusions

Organic manure application and use of artificial substrates for periphyton growth in Fingerponds result in increased fish yields. A maximum of 2,670 kg ha^{-1} was obtained from one pond in Gaba which provided 6.4 kg fish per person over a ten month growth period. Artificial substrates for periphyton provide additional natural food in the ponds and improve the diet of the fish. Fish production in Fingerponds can be limited by a) high recruitment and inbreeding between *O. niloticus* and *O. leucostictus*; b) falling water levels which lead to a deterioration in water quality c) fish species, e.g. *Oreochromis leucostictus* which results in a majority small sized fish between 100 and 200 g; d) high suspended solids and turbidity which lead to light limitation hence reduced primary productivity; and e) low zooplankton biomass. However, Fingerponds have been shown to be a promising semi-intensive, natural-based, fish-culture system for riparian communities around Lake Victoria.

Acknowledgements

This research was funded by the European Union, EU/INCODEV Fingerponds Project No. ICA4-CT-2001-10037 and in part by the Netherlands Government through the Netherlands Fellowship Programme through UNESCO–IHE, Institute for Water Education, Delft, The Netherlands. It was carried out in partnership with

Makerere University Institute of Environment and Natural Resources (MUIENR), Uganda. We are indebted to the management and staff of National Water and Sewerage Corporation and the National Fisheries Resources Research Institute of Uganda for providing technical support and facilities. We are forever grateful for the laboratory and field assistants in Kampala and Jinja who worked tirelessly.

References

APHA, 1992. Standard methods for examination of water and wastewater. 18th Edition. American Public Health Association. Washington, DC.

APHA, 1995. Standard methods for examination of water and wastewater, 19th Edition. American Public Health Association, Washington, DC.

Azim, M.E., Verdegem, M.C.J., Khatoon, H., Wahab, M.A., van Dam, A.A. & Beveridge, M.C.M. 2002. A comparison of fertilization feeding and three periphyton substrates for increasing fish production in freshwater pond aquaculture in Bangladesh. *Aquaculture* 212: 227-243.

Balirwa, J.S.1998. Lake Victoria wetlands and the ecology of the Nile Tilapia *Oreochromis niloticus* Linné. A.A. Balkema Publishers, Rotterdam, The Netherlands. 247 pp.

Bíro, P. 1995. Management of pond ecosystems and trophic webs. *Aquaculture* 129: 373-386.

Boyd, C.E. 1990. *Water quality in ponds for aquaculture*. Birmingham Publishing Co., Birmingham, Alabama. 482 pp.

Boyd, C.E. 1997. Practical aspects of chemistry in pond aquaculture. *Progress Fish-Cult.* 59: 85-93.

Brummett, R.E. 1995. Environmental regulation of sexual maturation and reproduction in tilapia. *Rev. Fish. Sci.* 3: 231-248.

Clover, J. 2003. Food security in Sub-Saharan Africa. *African Security Review* 12. 1: 5-15.

Colman, J.A. and Edwards, P. 1987. Feeding pathways and environmental constraints in waste-fed aquaculture: balance and optimization. In: Detritus and microbial ecology in aquaculture, Moriarty, D.J.W. & R.S.V. Pullin (Eds.), *ICLARM Conference Proceedings* 14: 240-281.

de Graaf, G.J., Galemoni, F., and Banzoussi, B. 1996. Recruitment control of Nile tilapia, *Oreochromis niloticus*, by the African catfish, *Clarias gariepinus* (Burchell 1822), and the African snakehead, *Ophiocephalus obscuris*. I. A biological analysis. *Aquaculture* 146: 85-100.

de Graaf, G.J. 2004. *Optimization of the pond rearing of Nile Tilapia (Oreochromis niloticus niloticus L.) The impact of stunting processes and recruitment control.* Ph.D thesis Wageningen University. 167 pp.

Delincé, G. 1992. *The ecology of the fish pond ecosystem with special reference to Africa.* Kluwer Academic Publishers, Dordrecht. 230 pp.

Denny, P. and Bailey, R.G. 1978. Survey of Nyumba ya Mungu reservoir, Tanzania. *Biological Journal of the Linnean Society* 10. 157 pp.

Dhawan, A. and Kaur, S. 2002. Pig Dung as Pond Manure. Effect on water quality, pond productivity and growth of carps in polyculture system. *NAGA, The ICLARM Quarterly* 25 (1): 11-14.

Duncan, A. 1975. Production and biomass of three species of Daphnia coexisting in London reservoirs. *Verh. Internat. Verein. Limnol.* 19: 2858-2867.

Edwards, P. 1999. Aquaculture and poverty: past, present and future prospects of impact. Discussion paper prepared for the 5th Fisheries Development Donors' Consultation held in Rome, *FAO.* 21-23 February 1999. Rome.

Egna, H.S. and Boyd, C.E. 1997. *Dynamics of pond aquaculture.* CRC Press, Boca Raton/New York. 437 pp.

Elhigzi, F.A.R., Haiderland, S.A. and Larsson, P. 1995. Interactions between Nile tilapia (O.n) and Cladocerans in ponds (Khartoum Sudan). *Hydrobiologia* 307: pp. 263-272.

Fernando, C.H. (ed.) 2002. *A guide to tropical freshwater zooplankton. Identification, ecology and impact on Fisheries.* Backhuys Publishers, Leiden, The Netherlands. 291 pp.

Glasser, F. and Oswald, M. 2001. High stocking densities reduce *Oreochromis niloticus* yield: model building to aid the optimization of production. *Aquatic Living Resources* 14: 319-326.

GOU, 2005a. Lake Victoria data base. Directorate of Water Development, Water Resources Management Department, Ministry of Water Land and Environment – The Republic of Uganda.

GOU, 2005b. Aquaculture technical manual. An introduction to commercial fish farming. 1. Ministry of Agriculture, Animal Industry and Fisheries. Government of Uganda. *National Agricultural Advisory Services (NAADS).* 65 pp.

Green, B.W., Phelcps, R.P. & Alvarenga, H. R. 1989. The effects of manures and chemical fertilizers on the production of *Oreochromis niloticus* in earthen ponds. *Aquaculture* 76: 37-42.

Hargreaves, J.A. and Huesel, L. 2000. A control system to stimulate diel pH fluctuation in eutrophic aquaculture ponds. *Journal of World Aquaculture Society* 31: 390-402.

Hem, S. and Avit, J.L.B. 1994. First results on 'acadja enclos' as an extensive aquaculture system (West Africa). *Bulletin of Marine Science* 55: 1040 -1051.

Hyslop, E.J. 1980. Stomach contents analysis-a review of methods and their application. *Journal Fish Biology* 17: 411-429.

Jalabert, B. and Zohar, Y. 1982. Reproduction physiology in cichlid fishes, with particular reference to *Tilapia* and *Sarotherodon.* pp. 129-140. In R.S.V. Pullin and R.H. Lowe-McConnell (eds.) *The biology and culture of tilapias.* ICLARM Conf. Proc. 7. 432 pp.

Jensen, A.L. 1986. Functional regression and correlation analysis. *Can. J. Fish. Aquat. Sci.* 43: 1742-1745.

John, D.M., Whitton, B.A. and Brook, A.J. (eds.). 2002. The freshwater algal flora of the British Isles. *An identification guide to freshwater and terrestrial algae.* Cambridge University press. 702 pp.

Jos, Á., Pichardo, S., Prieto, A.I., Repetto, G., Vázquez, C.M., Moreno, I. & Cameán, A.M. 2005. Toxic Cyanobacterial cells containing microcystins induce oxidative stress in exposed tilapia fish *(Oreochromis sp.)* under laboratory conditions. *Aquatic toxicology* 72: 261-271.

Kaggwa, R.C., Kansiime, F., Denny, P. and van Dam, A.A. 2005. A preliminary assessment of the aquaculture potential of two wetlands located in the northern shores of Lake Victoria, Uganda. In: J. Vymazal (eds.) *Natural and constructed*

Wetlands: Nutrients, metals and management. Backhuys Publishers, Leiden, The Netherlands. pp. 350-368.

Knud-Hansen, C.F. and Pautong, A. 1993. On the role of urea in pond fertilization. *Aquaculture* 114: 273-283.

Lorenzen, K. 2000. Population dynamics and management. In: M.C.M. Beveridge and B.J. McAndrew (eds.). *Tilapias: Biology and exploitation.* Kluwer Academic Publishers, Great Britain. pp. 163-225.

Lovshin, L.L. 1982. Tilapia hybridization. In: R.S.V. Pullin and R.H. Lowe-McConnell (eds.). The biology and culture of tilapia. *ICLARM* Conference Proceedings 7. 432 pp. Manila, Philippines. pp. 279-309.

Lowe-McConnell, R.H. 1979. Ecological aspects of seasonality in fishes of tropical waters. *Symp. Zool. Soc. Lond.* London. 44. pp. 219-241.

Lowe-McConnell, R.H. 1982. Tilapia in fish communities. In: R.S.V. Pullin and R.H. Lowe-McConnell (eds.). The biology and culture of tilapia. *ICLARM* Conference Proceedings 7. 432 pp. Manila, Philippines. pp. 83-113.

MacGrory, J. and Williams, D. 1996. Katha Fishing: Economics, Access patterns and potential conflicts with Fish Cage Culture, *CARE.* Bangladesh. 74 pp.

Machena, C. and Moehl, J. 2001. African Aquaculture: A Regional summary with emphasis on Sub-Saharan Africa. pp. 341-355. In: R.P. Subasinghe., P. Bueno., M.J. Phillips., C. Hough., S.E. McGladdery and J.E.Arthur (eds.). Aquaculture in the Third Millennium. Technical Proceedings of the Conference on Aquaculture in the Third Millennium, Bangkok, Thailand. 20-25 February 2000. NACA, Bangkok and FAO, Rome. *NACA/FAO*, 2001. 471 pp.

McGinty, A.S. 1985. Effects of predation by large mouth bass in fish production ponds stocked with *Tilapia niloticus. Aquaculture* 46: 269-274.

Milstein, A. 1993. Factor and canonical analyses: basic concepts, data requirements and recommended procedures. pp. 24-31 In: M. Prein, G. Hulata & D. Pauly (eds.) *Multivariate methods in aquaculture research: case studies of tilapias in experimental and commercial systems.* ICLARM Stud. Rev. 20. Manila, Philippines. 221 pp.

Moriarty, C.M. and Moriarty, D.J.W. 1973. Quantitative estimation of the daily ingestion of phytoplankton by *Tilapia nilotica* and *Haplochromis nigripinnis* in Lake George, Uganda. *J. Zool. Lond.* 171: 15-23.

Mosille, O.I.W. 1994. Phytoplankton species of Lake Victoria. In: Reports from the Haplochromis Ecology Survey Team (HEST). Report No. 30B. Zoologisch Laboratorium, Morphology Department, Leiden University, The Netherlands. 84 pp.

Nauwerck, A. 1963. Die Beziehungen zwischen Zooplankton und Phytoplankton imsec. *Erhen. Symb. Bot. UpSal.* 17(5). 163 pp.

Noakes, D.L.G. and Balon, E.K. 1982. Life histories of tilapias: an evolutionary perspective. In: R.S.V. Pullin and R.H. Lowe-McConnell (eds.). The biology and culture of tilapias. *ICLARM*, Manila. pp. 61-82.

Pechar, L., 1987. Use of an acetone: Methanol mixture for the determination of extraction and spectrophotometric determination of chlorophyll-*a* in phytoplankton. *Arch. Hydrobiol. Suppl.* 78: 99-117.

Pechar, L. 1995. Long-term changes in fish pond management as 'an unplanned ecosystem experiment': Importance of zooplankton structure nutrients and light for species composition of cyanobacterial blooms. *Water Science and Technology* 32: 187-196.

Pokornỳ, J. Přikryl, I., Faina, R., Kansiime, F., Kaggwa, R.C., Kipkemboi, J., Kitaka, N., Denny, P., Bailey, R., Lamtane, H.A. and Mgaya, Y.D. 2005. Will fish pond management principles from the temperate zone work in tropical Fingerponds. In: J. Vymazal. *Natural and Constructed Wetlands: Nutrients, Metals and Management.* Backhuys Publishers, Leiden, The Netherlands. pp. 382-399.

Ricker, W.E. (Ed.) 1971. *Methods for assessment of fish production in fresh waters.* IBP Handbook. No. 3 (2nd edition). International Biological Programme. Blackwell Scientific publications. Oxford and Edinburgh. 348 pp.

Ricker, W.E. 1975. Computation and interpretation of biological statistics of fish populations. *Bull. Fish. Res. Board. Can.*191. 382 pp.

Robotham, P.W.J. 1990. Trophic niche overlap of fry and juveniles of *Oreochromis leucostictus* (Teleostei, Chichlidae) in the littoral zone of a tropical lake (L. Naivasha, Kenya). *Revue Hydrobiologie Tropicale* 23: 209-218.

Teichert-Coddington, D.R. and Green, B.W. 1993. Tilapia yield improvement through maintenance of minimal oxygen concentrations in experimental grow-out ponds in Honduras. *Aquaculture* 118: 63-71.

Trewavas, E.T. 1983. Tilapiine Fishes of the genera *Sarotherodon, Oreochromis* and *Danakilia.* British Museum (Natural History) Publication, London UK. 583 pp.

Vollenweider, R.A. (Editor), Talling, J.F., and Westlake, D.F. 1974. *A manual on methods for measuring primary production in aquatic environments.* IBP Handbook No. 12. Second Ed. Int. Biological Programme, London. Blackwell Scientific Publications., Oxford. 225 pp.

Welcomme, R.L. (Compil.) 1979. The inland fisheries of Africa. CIFA Occasional paper 7. FAO. 69 pp.

Wetzel, R.G. 1983. *Limnology.* 2nd Edition. Saunders College Publishing. 860 pp.

Wetzel, R. G. and Likens, G. E. 1991. *Limnological analyses.* 2nd Edition. Springer-Verlag. 391 pp.

Wootton, R.J. 1990. *Ecology of teleost fishes. Fish and fisheries.* Series 1. Chapman and Hall, New York. 404 pp.

WFC, 2005. *Fish and food security in Africa.* World Fish Centre. Penang. Malaysia. 11 pp.

Wurts, W.A. and Durborow, R.M. 1992. Interactions of pH, carbon dioxide, alkalinity and hardness in fish ponds. *SRAC,* Publications No. 464. 4 pp.

Yakupitiyage, A. 1993. Constraints to the use of plant fodder as fish feed in tropical small-scale tilapia culture systems: an overview. In: S.J., Saushik., Luquet, P. (eds.). Fish nutrition in practice, Vol. 61. *Insitut. National de la Recherche Agronomique*, Les Colloques, Paris France. pp. 681-689.

Chapter 6

Parameterization and calibration of a model for fish production in seasonal wetland fish ponds ('Fingerponds')

Abstract

A dynamic model was developed to simulate fish growth in a seasonal wetland fish pond 'Fingerpond'. The model describes fish growth based on the natural food production. It incorporates key environmental and management factors: pond water depth, food availability, fish stocking densities, initial fish and fingerling weights at stocking, reproduction rate, manure type and application rates. The model was calibrated for two types of Fingerponds distinguished by differences in their hydrological regimes: ponds with fairly constant water levels (functional culture period 300 days); and those affected by inflowing rivers with decreasing water levels (200 days functional period). The stocked fish and fingerling were fed with natural organic manure (chicken) applied fortnightly at a rate of 1,042 kg ha^{-1}. The model was able to capture the dynamics of hydrology, nutrients and fish in the two types of Fingerponds demonstrating that similar fundamental processes underlie fish production in these systems. Model fish yields of up to 2,800 kg ha^{-1} were achievable. Water quality predictions such as chlorophyll *a* and dissolved inorganic nitrogen concentrations were comparable to field measurements. Using the model, nitrogen budgets for Fingerponds were calculated and quantitative estimates of all process flows were given. The model is preliminary and should be subjected to sensitivity analysis and validation before it can be used as a management tool. Main knowledge gaps pertain to light limitation of primary productivity and the food selectivity of the tilapia fish and their fingerlings. In its current state, the model is a research tool that identifies knowledge gaps and can be applied to frame hypotheses for further applied research. According to the outputs, the model can contribute to improving the management of Fingerponds.

Key words: Fingerponds, fish yield, dynamic model, manure, nitrogen budget, pond management

Publication based on Chapter 6: Kaggwa, R.C., van Dam, A.A., Kipkemboi, J. and Denny, P. Parameterization and calibration of a model for fish production in seasonal wetland fish ponds ('Fingerponds') Ecological Modelling (submitted).

Introduction

Fingerponds are earthen ponds dug at the landward edge of wetlands and stocked naturally with fish following flood recession. Fish trapped in these seasonal wetland fish ponds can be farmed and cropped. The ponds provide an alternative protein supply for resource-poor communities particularly in the drier seasons, as demonstrated recently in experimental Fingerponds in the Rufigi floodplain, Tanzania and in the wetlands surrounding Lake Victoria in Uganda and Kenya (Denny et al., 2006).

Fish production in fish ponds can be stimulated by enhancing natural food production through the application of organic (mainly animal) manures (Boyd, 1990; Delincé, 1992). Research on Fingerponds shows that manure application can enhance net primary productivity (NPP) to average levels of up to 2.9 g C m^{-2} d^{-1} (Chapter 3). This is higher than findings in Honduras (2.4 g C m^{-2} d^{-1}), but lower than Panama (4.4 g C m^{-2} d^{-1}) for organically fertilized ponds (Batterson et al., 1989; Green et al., 1989; Knud-Hansen et al., 1991a; Teichert-Coddington et al., 1992). The natural productivity in these systems is supported by an autotrophic pathway in which solar energy is used by primary producers (mainly phytoplankton) for the photosynthetic fixation of carbon and nutrients (notably nitrogen and phosphorus). The resulting algal biomass is utilized by fish, either directly or indirectly via the heterotrophic pathway in which heterotrophic organisms (bacteria, protozoa and invertebrates) decompose organic matter that can be utilized by fish (Schroeder 1978; Schroeder et al., 1990). The pathways are linked through fluxes of inorganic and organic nitrogen nutrients (Delincé, 1992).

Nitrogen plays a vital role in aquaculture, water quality management and fish growth due to its dual role as a nutrient and a toxicant. While nitrogen is deliberately added to pond systems to enhance primary productivity, some dissolved nitrogen components (notably unionized ammonia and nitrite) should be maintained below toxic levels by water exchange or filtering and recirculation (Lorenzen et al., 1997). Moreover, nitrogen is also an important determinant of the environmental impact of aquaculture ponds. Excess nitrogen may end up in the environment, causing eutrophication problems of surface waters. Several studies show that the greater part (64-89%) of all nitrogen added to a fish pond accumulates in the pond sediment or is lost through volatilization or discharge (Hargreaves, 1998). The challenge in fish pond management is therefore to maintain good water quality while maximizing the retention of nutrient inputs in harvestable products (Jiménez-Montealegre, 2001).

Fingerponds are semi-intensive integrated agriculture-aquaculture systems, unique in there dependence on natural flood events. Over a grow-out period, pond water levels are regulated by surface run-off, rainfall, evaporation and seepage. As the culture period progresses, water levels begin to fall and the pond may dry out completely. In such closed systems the interactions between pond inputs, water quality, primary productivity, natural food supply and fish growth need to be well understood to enable the formulation of management strategies. Regulation of fertilizer application regimes and rates is an essential part of the management strategy. If properly done, deterioration of pond water quality to levels unsuitable for fish survival and growth can be avoided while optimizing the food supply to the fish and avoiding negative impact on the environment.

Fish growth in Fingerponds is governed by the resultant external conditions (such as climate and water level) and internal pond factors (water quality, phytoplankton productivity, fish reproduction and growth). Most of these external and internal factors were investigated experimentally in field experiments (this thesis; Kipkemboi, 2006). To increase the understanding of this complex system and provide a consistent description of the Fingerpond production system, a modelling approach was adopted. A dynamic model was constructed which took into consideration the major factors determining fish growth in a Fingerpond. The objectives of the model were three-fold: a) to improve the understanding of Fingerponds by providing insight into the processes that determine fish growth; b) to relate fish yields to the natural food production and management of Fingerponds; and c) to compute nitrogen budgets for two different types of Fingerponds. The model was calibrated against data from four Fingerpond sites on the shores of Lake Victoria: two lake floodplain sites in Uganda; and two sites in Kenya also affected by inflowing river hydrology. For simplicity the Kenyan sites are referred to as 'river floodplain' sites.

Material and methods

Data collection
Datasets for construction and parameterization of the model were collected from field experiments over the period May 2003 to April 2004, and from the literature. A detailed description of the sites in Gaba and Walukuba (Uganda) and Kusa and Nyangera (Kenya) and of data collection and analytical procedures can be found in Chapters 2 and 5 (this thesis) and in Kipkemboi (2006). The following water quality data were collected monthly between 10.00 and 14.00 hrs over a full production cycle: temperature (^{o}C), dissolved oxygen (DO), total suspended solids (TSS), ammonium nitrogen (NH_4-N), nitrate nitrogen (NO_3-N), chlorophyll a (Chl a) and secchi depth visibility. Mean water quality values of the manured ponds in each location were used for model calibration. Pond water depth was measured daily.

Light extinction coefficients (k_t) were estimated from secchi depth measurements as described by Poole and Atkins (1929). Sediment samples were collected at the beginning and end of the culture period from the ponds and analyzed for total nitrogen. Daily measurements of photosynthetic active radiation (PAR) were taken in Kirinya wetland, Jinja (2 km from the Walukuba location) using a Skye 4 channel radiometer with light sensors (Skye Instruments Ltd. 21, Ddole Enterprise Park, Llandrindod Wells) and at the pond site in Kusa, Kenya using a Weather Hawk silicon pyranometer sensor (Campbell Scientific, Inc. West 1800 Logan, USA).

Total numbers and weights of adult fish (various species, but dominated by three of tilapia, *Oreochromis* species) and their fingerlings were determined at stocking and at harvest. The total nitrogen added to the ponds was determined from the manure input and from nitrogen fixed by Cyanobacteria. Additional data on tilapia growth and nitrogen cycling was adopted from literature on tilapia in fish ponds (Delincé, 1992; Hargreaves, 1998; Jiménez-Montealegre, 2001; Jamu & Piedrahita, 2002; Jiménez-Montealegre *et al.*, 2002; van Dam *et al.*, 2004; van Dam & Verdegem, 2005).

System description
Distinctive differences between the sites were noted for pond water levels. In Kusa, the ponds dried out completely and showed the most distinctive drop in water level (Figure 6.1a). Photosynthetic Activel radiation (PAR) daily variations differed greatly with notably higher values occurring in Kirinya wetland near Walukuba (Figure 6.1b).

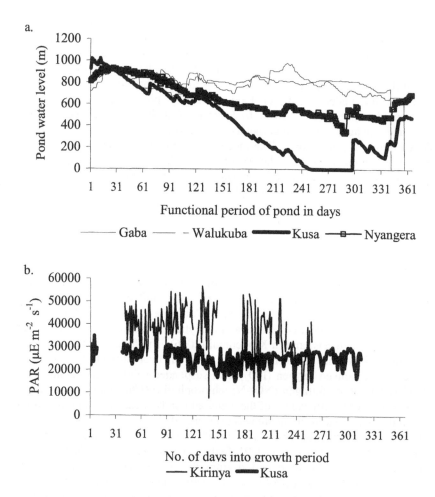

Figure 6.1. Daily pond water levels in the four fingerpond locations and Photosynthetic Active Radiation (PAR) in Kirinya near Walukuba and Kusa, May 2003 to April 2004.

Water quality in the two locations was also distinctively different (Table 6.1). In general, dissolved inorganic nitrogen (DIN) concentrations were lower in the lake floodplain ponds (Gaba and Walukuba) than in the river floodplain ponds (Kusa and Nyangera). However, DIN concentrations increased over the grow-out period particularly in Nyangera (Figure 6.2). Chlorophyll *a* concentrations did not show much variation with time with exception of Walukuba where a noticeable rise was noted in the last two months of the culture period. Suspended solids were relatively high (above 50 mg L^{-1}), especially in Nyangera. In general, secchi depth visibility in all the ponds was low though in Kusa it exceeded 25 cm.

Table 6.1. Water and fish variables for four fingerpond locations over a grow-out period

Water quality variables	Gaba		Walukuba		Kusa		Nyangera	
pH	7.7 -	8.7	8.6 -	9.3	7.5 -	8.2	8.5 -	8.9
EC (μS cm^{-1})	368 -	440	1194 -	1562	3592 -	6703	4300 -	8466
Temperature ($^{\circ}$C)	23.2 -	25.7	23.6 -	28.4	24.3 -	28.3	23.8 -	30.6
DO (mg L^{-1})	2.4 -	8.3	2 -	9.8	5.2 -	11.3	5.6 -	13.4
TSS (mg L^{-1})	65 -	262	83 -	197	60 -	140	135 -	35.3
DIN (mg L^{-1})	0.19 -	1.04	0.21 -	0.57	0.27 -	1.37	0.25 -	1.98
N:P ratio	3.2 -	9.7	2 -	4.3	2.6 -	99.6	3 -	14.1
Chl a (μg L^{-1})	23.6 -	131	41.9 -	500	21 -	120	15 -	145
secchi depth (cm)	15.1 -	19.6	12.8 -	21.3	24.7 -	39.3	11.9 -	16.3
Extinction coefficient K_t (m^{-1})	6.7 -	10.5	4.8 -	8.5	3.5 -	5.4	8.5 -	12.3
Fish variables	**Gaba**		**Walukuba**		**Kusa**		**Nyangera**	
Functional period (d)	310		219		228		189	
No of fingerlings initial census	151		18		613		1850	
No.of fingerlings final harvest	1218		1155		126		4459	
Individual fingerling weight initial census (g)	3.4		3.8		1.7		0.5	
Individual fingerling weight final harvest (g)	1.3		1.8		14.2		3.2	
Individual fish weight initial census (g)	36.8		41.8		87.8		30.3	
Individual fish weight final harvest (g)	55.5		46.8		85		120	
No. of fish initial census	113		138		46		150	
No. of fish final harvest	2095		1096		94		198	

Values presented are for manured ponds; two in Gaba and three each in Walukuba, Kusa and Nyangera. For water quality samples values represent ranges; minimum to maximum, whilst variables for fish are mean values.

Model development

Conceptual model and assumptions

The model relates fish yield to the productivity and management of an organically manured Fingerpond considering the major factors that limit fish growth: pond water levels, phytoplankton productivity as related to light and nutrients, and fish food production in relation to fish stocking density. Nitrogen was chosen as the model currency and all variables were expressed in g of nitrogen. In the model, three layers were distinguished: the pond sediment, which consists of soil with interstitial water; the detritus layer, where all dead organic material in the pond accumulates and is mineralized by microbes and other heterotrophytes; and the water column which harbours fish, phytoplankton and dissolved and suspended materials. Phytoplankton and detritus were regarded as the main food sources for fish growth (Figure 6.3). Additionally, pond hydrology determines the length of the culture period (Figure 6.1).

A number of simplifying assumptions were made. All fish that entered the pond were assumed to be tilapia (*Oreochromis spp.*). Periphyton and macrophytes as food sources for the fish were assumed not to play an important role and were excluded from the model. These can be included later if subsequent experimentation proves them to be significant. In general, zooplankton biomass in all ponds was low. Consequently, the initial zooplankton standing crop following flood recession was consumed immediately by the numerous fish fry leading to a low zooplankton production rate (this thesis). Zooplankton was therefore not included in the model.

Oxygen concentrations in the pond were assumed to be within acceptable ranges for growth and survival of tilapia; above 2 mg L^{-1} in the early morning and never dropping below 4 mg L^{-1} during the day (Chapters 3 and 5). Furthermore, it was assumed that no lethal abiotic factors existed in the ponds. It was also assumed that decomposition of organic matter in the pond supplied enough carbon dioxide to support phytoplankton growth (Boyd, 1972).

The protein content of the phytoplankton was assumed to be constant as well as the survival rate of fish. The pond was assumed to be homogeneously mixed. Denitrification was considered negligible during the culture period since there was hardly any nitrate in the ponds. Nitrogen fixation has been considered negligible in other studies (El Samara & Oláh, 1979; Lin *et al.*, 1988) but in this study, low N:P ratios (<7) were obtained. Coupled with high biomass densities of Cyanobacteria (including those with heterocysts like *Anabaena circinalis*), these conditions are likely to favour N fixation and a rate equation for N-fixation was added to the model, contributing to the phytoplankton biomass. All nitrogen was assumed to be in the form of organic N or in the form of dissolved inorganic nitrogen (DIN, the sum of ammonia and nitrate nitrogen; nitrite was below detectable limits).

Modules

The model was made up of three modules: hydrology, phytoplankton-water-detritus-sediment and fish (Figures 6.4a and 6.4b).

Hydrology module

Fingerponds are excavated in wetlands with the pond bottom situated below the water table. They are dependent entirely on natural flood events. Once the flood recedes pond water levels are regulated by seepage, precipitation, evaporation and runoff (Boyd, 1982; Yoo & Boyd, 1994). The hydrology module was based on the water balance model developed for Fingerponds in Kenya, (Kipkemboi, 2006). The hydrological equation used was computed from

$$\text{Gains} = \text{Losses} \pm \text{change in storage} \qquad \text{(Equation 6.1)}$$

where gains are constituted by precipitation (P), runoff (R) and seepage into the pond (S_i); and losses are seepage out (S_o) and evaporation (E) (assuming that there is no abstraction for irrigation). Changes in water depth (H) can therefore be described as:

$$dH = P + R + \left(S_i - S_o\right) - E \qquad \text{(Equation 6.2)}$$

Pond water levels were obtained from daily measurements of a staff gauge installed at the mid point of one length side of each pond. Evaporation was estimated from the Class 'A' pan evaporation using a pan coefficient of 0.81 according to Boyd (1985) after correction for rainfall on rainy days. Rainfall was determined daily at 09.00 hrs using a rain gauge stationed 5 m away from the second pond at each location. Groundwater levels were measured from five piezometers (wells) set along a transect at the second pond of each location; Wells 1 and 2 were positioned 10 and 2 m away from the deep end while wells 3, 4 and 5 were positioned 2, 10 and 20 m from the shallow end of the pond. All wells were located midway between the pond length. Tables with daily evaporation, precipitation and groundwater measurements for each site were included in the model as driving variables. Runoff was calculated from precipitation and the maximum infiltration rate (mm d^{-1}) using the curve number method developed by the U.S Soil Conservation Service (1972). A curve number of 90 was used for the wetland area based on a hydrological group D indicating clay

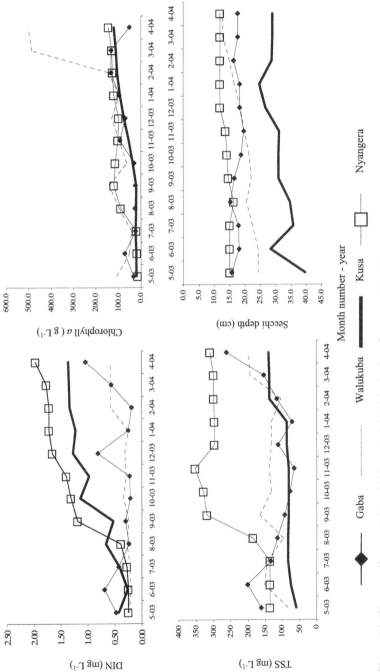

Figure 6.2. Water quality trends May 2003 to April 2004 in the four fingerpond locations.

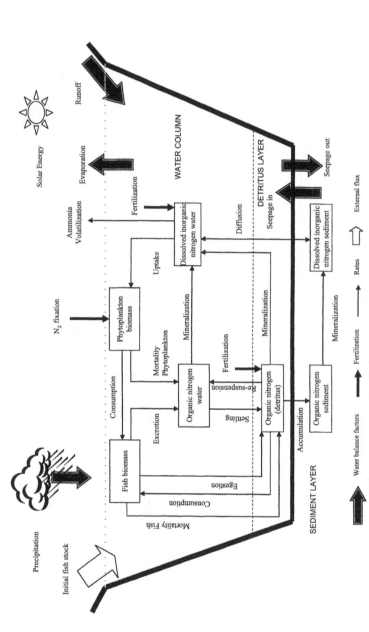

Figure 6.3. Conceptual model showing nitrogen forms as indicators for nutrient, energy flows and processes within a Fingerpond. The dotted line represents the division between the atmosphere and the water column, the dashed line the division between the water column and the detritus layer and the thick line the division between the detritus layer and the pond outline. The thick shaded blocked arrows represent the key water balance factors, the thick arrow the fertilization rate, the thin arrow the rates and the block un-shaded arrow the initial fish stock from the flood. Flows for denitrification are not included in the model. Fish represents adult and fingerlings.

soils. Seepage was calculated from measured groundwater levels using an empirical relationship between ground water level and seepage.

Phytoplankton-water-detritus-sediment module
This module consisted of the following state variables: phytoplankton biomass (N_{phyto}), fish biomass (N_{fish}), fingerling biomass (N_{fing}), dissolved inorganic nitrogen in the water columns (DINw), organic nitrogen in the water column (ONw), organic nitrogen in the detritus (ONdet), organic nitrogen in the sediment (ONsed), and dissolved inorganic nitrogen in the sediment (DINsed) (see Figure 6.4a).

The overall change in phytoplankton biomass was calculated as the difference between nitrogen uptake (limited by light and DIN concentration), mortality and consumption by fish. Phytoplankton growth rate was modelled as the difference between nitrogen uptake and mortality using the logistic equation:

$$\frac{dN_{phyto}}{dt} = r_{phyto} \times N_{phyto} \left(1 - \frac{N_{phyto}}{N_{phyto,max}} \right) \qquad \text{(Equation 6.3)}$$

or

$$\frac{dN_{phyto}}{dt} = r_{phyto} \times N_{phyto} - K_{Mor,Phy} \times N_{phyto}^2 \qquad \text{(Equation 6.4)}$$

in which N_{phyto} is the phytoplankton biomass (in g N), r_{phyto} is the maximum nitrogen uptake rate and $K_{Mor,Phy}$ is the mortality coefficient. Nitrogen uptake rate was limited by light and nitrogen using a Michaelis-Menten equation:

$$N \, uptake = r_{phyto} \times N_{phyto} \times \left(\frac{I_z}{I_z + K_{light}} \right) \times \left(\frac{C_{DINw}}{C_{DINw} + K_{DIN}} \right)$$
$$\text{(Equation 6.5)}$$

in which I_z is the light intensity (irradiance) at pond depth z ($\mu Em^{-2} \, s^{-1}$), K_{light} is the half saturation constant for light, C_{DINw} is the concentration of dissolved inorganic nitrogen in the water (in g m^{-3}) and K_{DIN} is the half saturation constant for DIN uptake (g m^{-3}). $I_z/(I_z+K_{light})$ and $C_{DINw}/(C_{DINw}+K_{DIN})$ are referred to as the light and nutrient limitation factor, respectively in the result section of this chapter. In the model, light intensity was computed based on the Lambert-Beer law from measured light intensity at the sites.

Dissolved inorganic nitrogen in the water column (DINw) was produced by mineralization of organic nitrogen in the water and detritus, diffusion from the sediment and from fertilization. Fertilization was computed from the manure application rate and the nitrogen content of the manure, assuming that a proportion of the manure contributed to DINw and the rest to ONdet. DINw disappeared through uptake by phytoplankton (Equation 6.5), through ammonia volatilization and through diffusion to the sediment. DIN (ammonia) volatilization was modelled as a first-order process:

$$DIN \, volatilization = k_{volat} \times C_{DINw} \qquad \text{(Equation 6.6)}$$

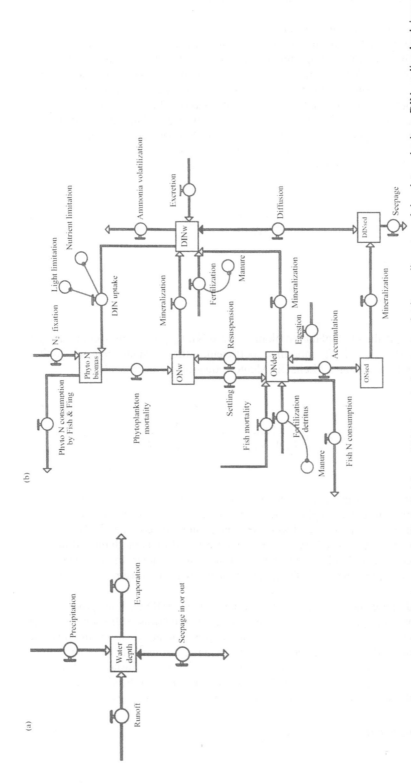

Figure 6.4a. Relational diagram for a) hydrology module and b) phytoplankton-water-detritus-sediment module. det = detritus, DIN = dissolved inorganic nitrogen, fing = fingerling, ON = organic nitrogen, phyto = phytoplankton, sed= sedimentation, w = water. Diagram modified from Stella version 8 (High Performance Systems, Inc., Hanover, USA)

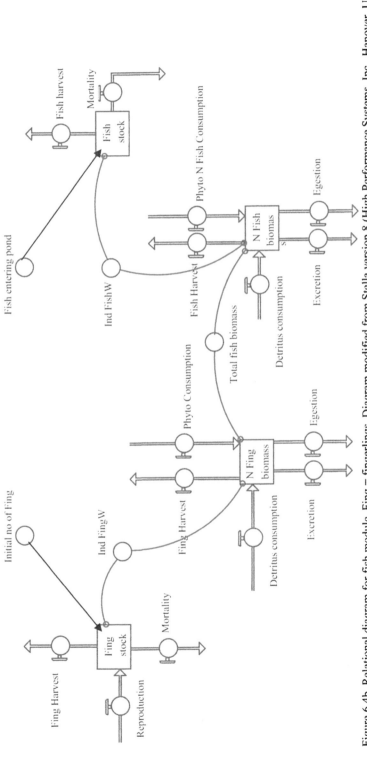

Figure 6.4b. Relational diagram for fish module. Fing = fingerlings. Diagram modified from Stella version 8 (High Performance Systems, Inc., Hanover, USA)

in which k_{volat} is the first-order rate constant for volatilization of ammonia nitrogen. Diffusion was modelled according to Fick's second law (Jiménez-Montealegre *et al.*, 2002; Jamu and Piedrahita, 2002):

$$DIN\ diffusion = K_{diff} \times P \times \left(\frac{C_{DINw} - C_{DINsed}}{D} \right) \times A \qquad \text{(Equation 6.7)}$$

in which K_{diff} is the diffusion coefficient for DIN (in $m^2 d^{-1}$), P is the porosity of the soil (dimensionless), C_{DINw} and C_{DINsed} are the concentrations of DIN in water and soil, respectively, D is the sediment depth (m) and A is the pond surface area (m^2).

Changes in organic nitrogen in the water column (ONw) were calculated as the difference between phytoplankton mortality (contributing dead algal matter to the ONw; see equation 6.4), re-suspension of organic nitrogen in the detritus (ONdet), settling of ONw to the detritus layer at the pond bottom and mineralization of ONw to DINw. All these rates were modelled as first-order equations (Equation 6.6).

Changes in organic nitrogen in the detritus (ONdet) were computed as the difference between settling, fish mortality, fertilization and egestion by fish (all contributing to production of ONdet), and re-suspension, consumption by fish, accumulation of organic N in the sediment and mineralization to DINw (all contributing to disappearance of ONdet). For a description of fish mortality, egestion and consumption see fish module section. Mineralization and accumulation were described as first-order processes.

Organic nitrogen in the sediment (ONsed) was modelled as the balance of accumulation from ONdet and mineralization to DIN in the sediment pore water (DINsed), both as first-order processes. Dissolved inorganic nitrogen in the pore water (DINsed) was assumed to diffuse to and from the water column. Loss of DIN from the pore water through seepage was calculated from the seepage rate in the hydrology module and the concentration of DINsed.

Fish module
In the fish module, a distinction was made between fish entering the pond with the flood, and fingerlings originating from fish reproducing within the ponds after the start of the culture period (Figure 6.4b). The fish module consisted of the state variables fish biomass (N_{fish}, in g nitrogen), fingerling biomass (N_{fing}, in g nitrogen), number of fish (Stk_{fish}) and number of fingerlings (Stk_{fing}).

For the number of fish, it was assumed that after initial stocking of the fish by the flood, the number of fish decreased according to a first-order equation. From the total number of fish, the number of mature fish was calculated by assuming that the individual fish weight in the population was normally distributed with an average weight W_{fish} and a standard deviation SD_{fish}. If the weight at first maturity was $W_{fish,mat}$, then the normal deviate Z was calculated as:

$$Z = \frac{\text{abs}(W_{fish,mat} - W_{fish})}{SD_{fish}} \qquad \text{(Equation 6.8)}$$

From Z, the area under the normal curve can be derived (see any statistics textbook; e.g., Snedecor and Cochran, 1980) and the proportion of immature fish ($Prop_{NotMat}$) was calculated as follows:

$$\text{Prop}_{\text{NotMat}} = \text{IF}\left(W_{fish} < W_{fish,mat}\right) \text{THEN} \left(Anc + 0.5\right) \text{ELSE} \left(0.5 - Anc\right)$$

(Equation 6.9)

in which Anc is the area under the normal curve. The number of immature fish was then calculated by multiplying with the total number of fish.

Changes in fish biomass (Nfish) were modelled as the difference between the consumption of phytoplankton and detritus; and egestion, excretion and mortality (van Dam & Verdegem, 2005). Feed intake by an individual fish was expressed with a logistic equation:

$$\text{Feed intake} = r_{fish} \cdot W_{fish} \cdot \left(1 - \frac{W_{fish}}{W_{fish,max}}\right)$$

(Equation 6.10)

in which r_{fish} is the dry matter feed intake of the fish (d^{-1}), W_{fish} is the average individual fresh weight of the fish (g), and $W_{fish,max}$ is the maximum individual weight of the fish in the population (g). The model computes W_{fish} by dividing the total fish biomass by the number of fish. Total consumption of the fish stock (in g N d^{-1}) was expressed as:

$$\text{N cons fish} = r_{fish} \times W_{fish} \times \left(1 - \frac{W_{fish}}{W_{fish,max}}\right) \times Stk_{fish} \times C_{feedN}$$

(Equation 6.11)

in which C_{feedN} is the nitrogen content of the feed (% dry matter, dm).

In the model, the total consumption was divided among the phytoplankton and the detritus. It was assumed that tilapia have a preference for phytoplankton and switch to detritus when the concentration of phytoplankton drops to lower levels. This was achieved by relating the proportion of phytoplankton in the diet ($\text{Prop}_{\text{PhytoDiet}}$) to the concentration of Chorophyll a using a Michaelis-Menten equation:

$$\text{Prop}_{\text{Phyto,Diet}} = \text{Prop}_{\text{Phyto,Diet,max}} \times \frac{C_{chla}}{C_{chla} + K_{chla,fish}}$$

(Equation 6.12)

in which C_{chla} is the concentration of Chlorophyll a and $K_{chla,fish}$ is the half saturation constant for phytoplankton uptake by fish. The proportion of detritus in the diet is then calculated as

$$\text{Prop}_{\text{DetrDiet}} = 1 - \text{Prop}_{\text{PhytoDiet}}$$

(Equation 6.13)

Egestion and excretion were computed from the consumption rate assuming that the assimilation percentage (digestibility) of the nitrogen was constant and that a fixed percentage of the assimilated nitrogen was excreted:

$$N\, egestion = N\, cons\, fish \times \left(1 - DC\right)$$

(Equation 6.14)

$$N\ excretion = N\ cons\ fish \times DC \times Excr \qquad \text{(Equation 6.15)}$$

in which DC is the nitrogen digestibility coefficient and Excr is the proportion of digested nitrogen that is excreted.

The number of fingerlings (Stk_{fing}) was computed from the difference between the reproduction rate and the fingerling mortality rate. Reproduction rate was formulated as:

$$\text{Reprod rate} = \frac{Stk_{fish,mat} \times Fec \times SexRat}{SpawnInt} \qquad \text{(Equation 6.16)}$$

in which Fec is the fecundity (no. of surviving fry per female per spawn), SexRat is the sex ratio of the fish (expressed as the number of females divided by the total number of fish) and SpawnInt is the spawning interval (d).

Fingerling growth rate, egestion and excretion were modelled identically to fish biomass and growth. However, fingerling individual weight was assumed to be constant. This meant that any change in fingerling biomass was a result of changes in the numbers of fingerlings.

Model implementation and parameterization

The model was implemented using Stella version 8 (High Performance Systems, Inc., Hanover, USA) with rectangular integration and a time step of 1/32 day. Parameter values were derived from various sources. From our own field observations and measurements, data on pond hydrology, light extinction coefficient, dissolved inorganic nitrogen, chlorophyll *a* concentrations and fish growth and yield were derived. Other parameter values were obtained by calibration of the model or from literature on similar models of aquaculture systems (mainly Jamu, 1998; Jiménez-Montealegre 2001; Jiménez-Montealegre *et al.*, 2002; van Dam & Verdegem, 2005). Parameter values uses are summarized in Table 6.2.

The model was initialized using data from the experiments as shown in Table 6.3. Two versions of the model were used. The first version represented a lake floodplain pond with a culture period of 300 days, while the second version represented the river floodplain pond with a decreasing water level and a culture period of 200 days. For both versions, the relationship between groundwater level and seepage was calibrated to the actual water levels in Gaba and Kusa (lake and river floodplain pond, respectively). Further calibration was then done by adjusting the uptake rates of nitrogen by phytoplankton and the food consumption rates of fish to give levels of dissolved inorganic nitrogen, chlorophyll *a* and fish that resembled the observations in the experimental sites as much as possible. Manure input rate was 20 kg per 2 weeks per pond, manure N content 2.3% in dry matter, initial fish density 1 fish per m^2, sex ratio of 1:1, and no harvesting was done during the culture period.

Table 6.2. Parameters used in the model

Process	Parameter	Name	Unit	Value	Remarks
Hydrology	maximum filtration rate	Max FR	mm d⁻¹	28.2¹, 25.1²	
	critical depth (mm)	Crit depth	mm	300	
Phytoplankton DIN uptake	maximum phytoplankton DIN uptake rate	r_{phyto}	d⁻¹	2	by calibration
	half saturation constant DIN limitation	K_{DIN}	g m⁻³	0.7	Chen & Orlob, 1975
	half saturation constant light limitation	$K_{light\ lim}$	dL	200	
	extinction coefficient non algal	$K_{non\ algal}$	dL	8¹, 5²	
Consumption by fish (phytoplankton)	fish dry matter total feeding level	Fdm FL	dL	0.009	by calibration
	fingerling dry matter total feeding level	Fing dm FL	dL	0.048	by calibration
	half saturation constant Chla in diet (fish)	$K_{Chla\ fish\ diet}$	dL	50	
	half saturation constant Chla in diet (fingerling)	$K_{Chla\ fish\ diet}$	dL	0.5	
	maximum proportion phytoplankton in fish diet	max prop. phyto in fish diet	dL	0.75	
	maximum proportion phytoplankton in fingerling diet	max prop. phyto in fing diet	dL	0.5	
Egestion	digestibility coefficient	DC	dL	0.5	Gangadhar et al., 2004
Excretion	nitrogen excretion rate as proportion of digested nitrogen	N excretion	dL	0.3	Oh et al, 2001
Mineralization	first-order mineralization rate constant detritus	$K_{min\ det}$	g N d⁻¹	0.01	van dam & Verdegem, 2005
	first-order mineralization rate constant sediment	$K_{min\ sed}$	g N d⁻¹	0.005	van dam & Verdegem, 2005
Diffusion	diffusion coefficient for DIN	K_{diff}	m² d⁻¹	0.002	Jamu, 1998; Jiménez Montealegre, 2001
	soil porosity	P	dL	0.8	
	sediment depth	D	m	0.15	
Mortality of Phytoplankton	mortality rate constant (phytoplankton)	$K_{mort,\ phyto}$	g N d⁻¹	0.005	Jamu, 1998; Jiménez Montealegre, 2001
	mortality rate constant (fish)	$K_{mort,fish}$	g N d⁻¹	0.0006	
	mortality rate constant (fingerling)	$K_{fing,mort}$	g N d⁻¹	0.015	
Reproduction	fecundity	Fe	dL	25	by calibration
	spawn interval	SpawnInt	d	30	
	sex ratio	SexRat	dL	0.5	
Volatilization	first-order ammonia volatilization coefficient	K_{volat}	g N d⁻¹	0.019	by calibration
Settling	first-order settling rate constant	K_{sett}	g N d⁻¹	0.005	by calibration
Resuspension	first-order resuspension rate constant	Kresus	g N d⁻¹	0.0025	by calibration
Accumulation	first-order sediment accumulation coefficient	K_{accum}	g N d⁻¹	0.005145	by calibration
Nitrogen fixation	Nitrogen fixation coefficient	K_{Nfix}	g N d⁻¹	0.0009	by calibration
General parameters	phytoplankton Chlorophyll a content	Phyto Chl a content	g Chl a (g dm)⁻¹	0.01	
	critical dissolved inorganic nitrogen in water	Crit DINw	g m⁻³	0.1	Delincé, 1992
	detritus density	Initial det density	g N m⁻³	0.3	
	detritus N content	Det N content	g N	0.04	
	fish weight at first maturity	Wrmat	g	50	
	standard deviation fish weight	Sigma W	dl	15	
	weight at infinity	Winf	g	500	
	irradiance standard deviation	Irrad sd.	dL	280	
	phytoplankton N content	Phyto N content	g N (g dm)⁻¹	0.08	van Dam & Verdegem, 2005

Unless indicated values were based on estimates from own observations. dl denotes dimensionless, ¹ lake floodplain ponds and ² river floodplain ponds, ** value computed from ranges given.

Table 6.3. Initial conditions of simulations

Variable	Symbol	Unit	Value
Organic Nitrogen in water dry matter density	ONw dm density	g N dm m^{-3}	0.5
Detritus density	Det density	g N dm m^{-3}	0.3
Dissolved inorganic nitrogen in water	C$_{DINw}$	g N m^{-3}	0.471[1], 0.570[2]
Individual fingerling stocking weight	FingW	g	1
Individual stocking fish weight	FishW	g	30
Number of fingerling stocking pond	Fing no.	dl	200
Organic nitrogen in water	ONsed	g N m^{-3}	3
Phytoplankton chlorophyll a	Initial phyto Chl$a$	µg L^{-1}	34[1], 50[2]
Pond water depth	Initial pondw depth	mm	720

dl denotes dimensionless

Results

Model simulations

Hydrology
The model was calibrated to simulate the water level in the two types of Fingerponds using the empirical relationship between seepage and ground water levels. For both pond types, a linear relationship between groundwater level and seepage was used, with low groundwater levels leading to seepage out of, and high groundwater levels to seepage into the pond. For the lake floodplain pond, the pond did not empty whilst for the river floodplain pond, the water levels began to drop after 20 days reaching the critical pond water depth (300 mm) by the end of the culture period (Figure 6.5). The lake pond had higher and more constant precipitation and lower evaporation than the river floodplain pond leading to a more constant water level. Seepage rates were on average higher in the lake floodplain pond.

Phytoplankton, water, detritus and sediment
In the lake floodplain pond, simulated Chl a in the water column (Chl$a$w) increased steadily to about 100 g m^{-3} after 150 days and remained within this level until the end of the culture period. This was similar to what was measured in these ponds (Figure 6.6a). Dissolved inorganic nitrogen in the water column (DINw) increased to about 6.5 g m^{-3} on day 250, after which it declined somewhat. This was much higher than the constant level of around 0.5 g m^{-3} that was measured (Figure 6.6b). The river floodplain pond showed increasing chlorophyll a levels to about 350 g m^{-3} at the end of the culture period (measured: constant between day 25 and 100) and slowly increasing DINw concentrations between 0.25 and 0.85 g m^{-3} (similar to measurements; Figure 6.6b).

The model suggested increasing light limitation in the lake pond, with the light limitation factor decreasing from about 0.60 to 0.44 during the culture period. Nutrient limitation decreased in this pond, as shown by the increase in nutrient limitation factor up to a value of 0.9 at the end of the culture period (Figure 6.7). In the river pond, light limitation was less severe than in the lake pond with values of

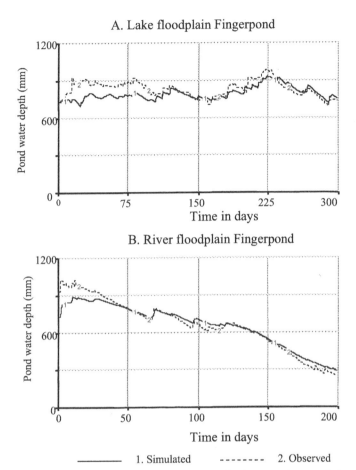

Figure 6.5. Simulated and observed pond water depth over culture period for A. lake and B. river floodplain fingerpond, baseline scenario.

the limitation factor between 0.75 and 0.88. Nutrient limitation in the river pond was strong in the beginning (0.10) but decreased towards the end of the culture period (0.50).

In both pond types, the model showed an increase in organic nitrogen in the water (to about 9 and 35 g N m^{-3} in the lake and river pond, respectively) and detritus (to about 18 g N m^{-3} at the end of the period) (Figure 6.7).

Fish
The total number of stocked fish reduced steadily in both pond types due to mortality (Figure 6.8). In both ponds, all fish were mature after about 170 days. The number of fingerlings increased to 4412 and 3901 at the end of the culture period in the lake and river pond, respectively. Individual weight at the end of the culture period of the stocked fish was 175 g (lake pond) and 108 g (river pond). Total fish yields were 2857 kg ha^{-1} and 1659 kg ha^{-1}, respectively. In the lake pond, this yield consisted of roughly equal biomass of fish and fingerlings (about 27000 g each). In the river pond, there were more adult fish (18000 g) than fingerlings (13000 g).

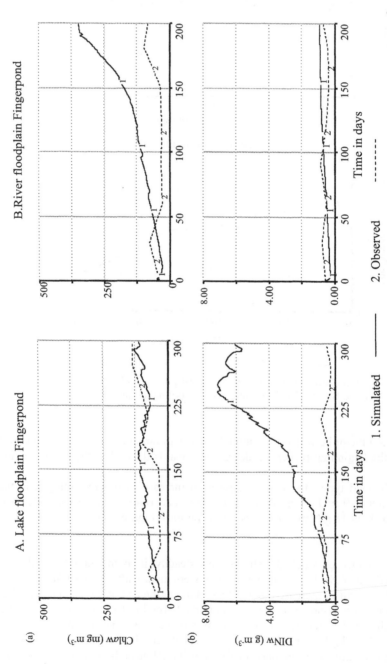

Figure 6.6. Simulated and observed results for Chlorophyll a (Chl a, in mg m^{-3}) and dissolved inorganic nitrogen (DIN, in g m^{-3}) in water column.

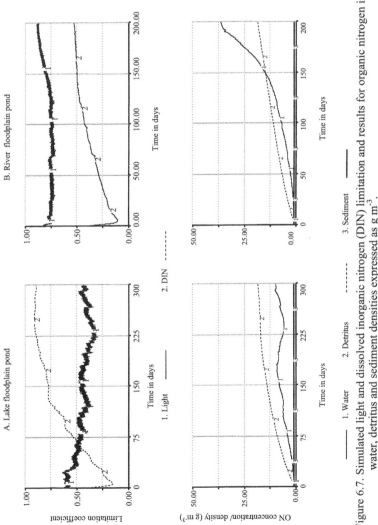

Figure 6.7. Simulated light and dissolved inorganic nitrogen (DIN) limitation and results for organic nitrogen in water, detritus and sediment densities expressed as g m^{-3}.

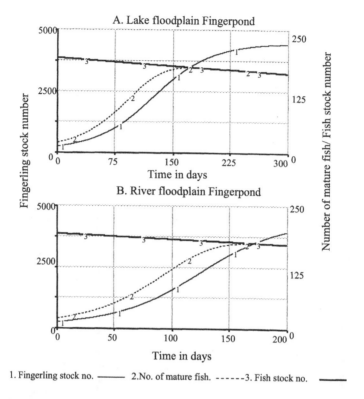

Figure 6.8. Simulated fingerling and fish number

The proportion of phytoplankton in the diet of the adult fish and fingerlings was about 0.5 and 0.33 in the lake pond for most of the culture period. In the river pond, the proportion of phytoplankton in the diet increased during the culture period from 0.25 to 0.65 for the adult fish and from 0.24 to 0.44 for the fingerlings. The total uptake of nitrogen by fingerlings at the end of the culture period was about double that of the adult fish in the lake pond. In the river pond, fingerlings ate even more than twice the amount of nitrogen than the adult fish (Figure 6.9).

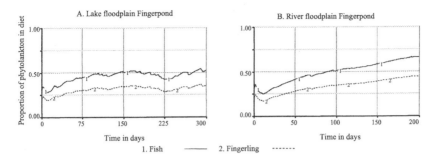

Figure 6.9. Proportion of phytoplankton in fish and fingerling diet

Nitrogen budgets and flows

Table 6.4 summarizes the nitrogen budgets for both pond types. Total nitrogen input was 9882 and 6590 g for the lake and river pond, respectively. The majority of this input came from the manure application, of which 95% was assumed to contribute directly to the organic nitrogen in the detritus. Less than 1% of the input came from nitrogen fixation, resulting in a flow of about 0.5 mg N m^{-2} d^{-1} in both ponds (Table 6.5). This is very low compared to values of 6-57 mg N m^{-2} d^{-1} for nitrogen fixation reported in the literature (Schroeder, 1987; Lin et al., 1988; Acosta Nassar et al., 1994).

The major part of N input accumulated in detrital organic nitrogen at the pond bottom (35% and 53% for the lake and river pond, respectively) and in the water (14% and 32%, respectively). The lower accumulation in the lake pond was related to the higher simulated DIN concentration in this pond, which led to much higher volatilization losses with the current model settings (25% of N input in the lake pond, against only 4% in the river pond). This is directly related to the stronger light limitation in the lake pond, which also resulted in lower uptake of N by phytoplankton (0.84% of N input, against 1.82% in the river pond). Volatilization rates reported in the literature range from 0.5 to about 50 mg N m^{-2} d^{-1} (Schroeder, 1987; Abdallah et al., 1996; Gross et al., 2000) and the rates calculated by the model fall within this range (44.6 and 6.3 mg N m^{-2} d^{-1}). Loss of DIN through seepage was higher in the lake pond (4% of N input) than in the river pond (0.4% of input).

DIN uptake by phytoplankton in the models was 295 and 357 mg N m^{-2} d^{-1}, which is within the range reported for temperate fishponds (150-450 mg N m^{-2} d^{-1}) but lower than rates reported for productive tropical ponds (750-1500 mg N m^{-2} d^{-1}; Hargreaves, 1998). Of the total N input, 12.7% and 10.0% ended up in fish biomass in the lake and river model, respectively, of which slightly more than half was in fingerling biomass.

Table 6.4. Nitrogen balance for the lake and river ponds

| Pond type | Lake pond (300 days) | | River pond (200 days) | |
Variable	g N	% of total input	g N	% of total input
total ON fertilization	9364	94.76	6243	94.73
total DIN fertilization	493	4.99	329	4.99
total N fixation	25	0.26	18	0.28
total N input	9882		6590	
Phytoplankton	83	0.84	120	1.82
DINw	678	6.86	-16	-0.24
DINsed	77	0.78	9	0.14
ONw	1358	13.74	2078	31.53
ONdet	3461	35.03	3470	52.66
ONsed	3	0.03	2	0.03
fish biomass	572	5.79	323	4.91
fingerling biomass	681	6.89	336	5.09
ammonia volatilization	2571	26.01	242	3.68
DIN seepage	398	4.02	26	0.39

Table 6.5. Average nitrogen flows (in mg N m^{-2} d^{-1}) for a lake and river pond

Process	Lake pond	River pond
DIN uptake by phytoplankton	294.64	356.59
Consumption		
of phytoplankton by fish	20.00	18.08
of phytoplankton by fingerling	28.59	23.30
of detritus by fish	11.74	8.44
of detritus by fingerling	31.20	19.44
Egestion		
by fish	15.87	13.26
by fingerling	29.89	21.37
Excretion		
by fish	4.76	3.98
by fingerling	8.97	6.41
Mortality		
of fish	1.17	0.86
of fingerling	9.10	6.22
of phytoplankton	245.05	312.56
Diffusion		
into sediment	6.59	0.00
into water column	0.19	0.92
Mineralization		
in water	213.47	240.63
from detritus	121.69	101.98
from sediment	1.84	1.83
Resuspension	30.42	25.49
Settling	38.43	43.31
Accumulation	1.89	1.88
Ammonia volatilization	44.63	6.31
DIN seepage	6.90	0.67
Nitrogen fixation	0.44	0.48
N input from fertilizer		
ON into detritus	162.57	162.57
DIN into water	8.56	8.56
Maximum NPP (in g Cm^{-2}d^{-1})	1.95	2.36

Discussion

Light, nutrient and food limitations

Fingerponds are turbid, and one of the main differences between the lake and river floodplain ponds was the higher turbidity in the lake floodplain pond model. This was based on observations in the Gaba site (Uganda), where high non-algal turbidity was observed. However we cannot conclusively attribute this to the difference in pond type but rather suggest that it is more a site specific problem. The difference was incorporated in the model through the non-algal turbidity coefficient which was 8 and 5 for lake and floodplain ponds, respectively (Table 6.2). As a result, light limitation was stronger in the lake ponds. On the other hand, nutrient limitation was stronger in the river ponds because DIN concentrations remained low while they increased in the lake ponds. Although it is hard to verify whether the light and nutrient limitation in the model mimicked the actual limitations occurring in the pond, the resulting overall flows of 295 and 357 g N m^{-2} d^{-1} (converted to 1.95 and 2.37 g C m^{-2} d^{-1} using the Redfield ratio) come close to what was measured in the Fingerponds (Chapter 3).

A preliminary sensitivity analysis with increasing manure applications suggested that applications over 30 kg per fortnight did not further increase chlorophyll concentrations much and did not have much effect on fish weight. Nutrient limitation in the ponds is apparent, and with low N:P ratios more so for nitrogen. At higher manure application rates, DIN limitation reduces especially towards the end of the culture period signifying the concentration of nutrients in low water levels. The quantity of N added to aquaculture ponds by fixation depends largely upon species composition of the phytoplankton community, significant proportion of heterocystous cyanobacteria and ammonia concentration. For this study we were not able to quantify the proportion of heterocystous cyanobacteria in the pond. With high ammonia concentrations, N$_2$ fixation is minimal (Lin et al., 1988). In the current model, N$_2$ fixation does not contribute significantly to nitrogen input into the pond. In view of the low N:P ratios and the presence of Cyanobacteria, N-fixation may be more important than the model suggests. N losses through leaching were low, even in the river floodplain pond because water level decrease was due more to evaporation than to seepage. DIN limitation can be lowered with improved N input but this effect is limited by turbidity-induced light limitation. Mitigation of turbidity should be a priority for improving phytoplankton productivity in Fingerponds. This will also increase the success of periphyton substrates in Fingerponds (Chapter 4).

The high number of fingerlings exerts a high feeding pressure. The model suggests that as the culture period progresses, their nitrogen consumption rates are at least twice as high as the consumption rates of the adult fish. This stresses the need to manage the stocking density, which is quite a challenge since stocking densities in fingerponds are dependent on natural flood events. Stock management strategies have been derived such as manual sexing, removal of fry and fingerling using traditional fishing methods, e.g. use of baskets and netting (Brummett, 1995; 2002). From the study in Chapter 5, we see that curbing recruitment in such systems is very difficult. The application of these methods to the Fingerpond systems may in themselves present problems particularly in the more resource-poor communities. In principle, the model can be used to evaluate the effects of intermittent harvesting and sex ratio manipulation (e.g., through manual sexing). However, apart from the population dynamics, more data would also be needed on the effects of density changes on competition for food resources among fish of different size within one species. Modelling a pond with more than one species would present even more challenges.

Food preference of fish

Food and feeding habits of fish vary between fish species, and within species with respect to size, life history stage, type of food available, season, time of day and locality in which they are found. For adult tilapia, their preference depends more on fish species. Some tilapia are microphytophagous feeding on phytoplankton. Others like *Tilapia rendalli* (Boulenger) are macrophytophagous (Chifamba, 1990). Adult *Oreochromis* spp. feed on phytoplankton, detritus or periphyton (Jauncey & Ross, 1982; Getachew, 1987; Stickney, 1996). The juveniles are omnivores, feeding on algae, zooplankton and insect larvae (Tudorancea *et al.*, 1988). Fish fry and fingerlings require more protein in their diet (45-50 % protein), hence their preference for zooplankton. An ontogenetic shift from zooplanktivory to phytoplanktivory can occur depending on the age of the fish (Trewavas, 1983; Robotham, 1990).

In Fingerponds, zooplankton colonization of the ponds following natural stocking of ponds is low and zooplankton biomass densities do not develop to adequate levels over the culture period. The high number of fingerlings exerts a high grazing pressure on the zooplankton, hence the low biomasses and subsequent low productivity. It is possible that the fish prey on the larger zooplankton, leaving the smaller ones which cannot develop a high biomass and therefore have no capacity for effective grazing on algae and larger bacteria (Porkornỳ *et al.*, 2005). Based on the apparent absence of zooplankton from Fingerponds, it was decided not to incorporate zooplankton in the model. If zooplankton is important in the Fingerponds food web, they would be competing for the phytoplankton and detritus with fish and fingerlings. Furthermore, this would present an extra trophic level between primary productivity and fish. Therefore it is questionable if the presence of zooplankton would indeed make fish production in Fingerponds higher or more efficient.

Tilapias are opportunistic and quite versatile in their choice of food. In the event that one type of food is missing they are able to shift to other available food resources. Tadesse (1999) reported a diet comprising of algal-based detritus, macrophyte scraps, phytoplankton and silt in adult *O. niloticus* in a lake. If phytoplankton is not enough they resort to detritus or periphyton. From the stomach contents of the fish (Chapter 5), it is clear that detritus constitutes the bulk of the fish diet. In this case, the model is realistic. Simulations with higher proportions of phytoplankton in the fingerling diet showed that phytoplankton runs out at the end of the culture period. In the model, detritus is less nutritious than phytoplankton and does not support the rapid growth of fish. In reality, detritus is a mix of low quality organic matter and more nutritious heterotrophic microorganisms and macroinvertebrates. No data on the dynamics of heterotrophic food sources for the fish is available, and the model needs to be improved in this area.

The model is based on the assumption that only *Oreochromis* species stock Fingerponds. The initial censuses in various Fingerponds showed that several other fish species were present, such as haplochromines, clariids and *Protopterus aethiopicus* though in smaller numbers (Porkornỳ *et al.*, 2005). Inclusion of these other fish species in the model should create a more realistic situation but modelling their feeding behaviour (selectivity, competition) would be very complex. Presence of piscivorous fish, e.g. *Clarias* can be used as a means of reducing fingerlings and fry. This would reduce the competition for phytoplankton, making it more readily available for the bigger fish. This may result in increased fish production.

Nitrogen budgets and flows

Total fish N biomass accounted for 10-12 % of the total N input. This is in the same range as that found in other fish ponds with manure as an input (Azim, 2001;

Edwards, 1993; Acosta-Nassar *et al.*, 1994; Gross *et al.*, 2000; Hargreaves, 1998). About half of the N in fish was contained in fingerling biomass, again stressing the importance of fish reproduction in the functioning of the ponds. The organic nitrogen input of up to 244 mg N m^{-2} d^{-1} from manure in the fingerponds is lower than the normal range for other semi intensive fish ponds of 700 to 800 mg m^{-2} d^{-1} (Schroeder *et al.*, 1990; Knud-Hansen *et al.*, 1991b). For Fingerponds, pond water levels are a key driving force in how much manure can actually be added to such closed systems. In tropical ponds, N uptake ranges from about 750 to 1500 mg N m-2 d^{-1} have been observed (Hargreaves, 1998) which is higher than the maximum of 249 mg m^{-2} d^{-1} obtained in our ponds. This demonstrates the limitation by DIN in these ponds.

Sedimentation and resuspension play a major role in fingerponds, further enhanced by the high pond turbidity. In ponds fertilized by manures or agricultural by-products, direct consumption by fish is minimal and most of the input settles to the sediment. In addition, as much as 50% of the algal standing crop may settle to the sediment surface each day (Schroeder *et al.*, 1991). The model does not account for increased resuspension as a result of wind action. This may be an important effect in Fingerponds, especially as water levels are reduced in the course of the culture period and the model may be improved in this area.

Organic nitrogen accumulation in detritus and sediment accounted for 35-53 % of the total N in the pond systems. This is typical of other fishponds as pointed out by Hargreaves (1998). In the model, most of this accumulation was in the detrital layer but no data on the distribution of organic nitrogen among the organic detritus and the firm sediment were available. Although the accumulation demonstrates the importance of the detrital pathway for N cycling in fish ponds (including Fingerponds), there is a need for more research into the processes governing this pathway. Sampling techniques for pond sediments often make no distinction between the soft detritus and the hard sediment. The accumulation of nitrogen in the sediment emphasizes the potential of the pond sediments as fertilizers for vegetables and other crops in integrated aquaculture systems.

Model performance, improvement and application

The model has been developed with data from two sets of ponds only, in Uganda and Kenya. Further experimentation with a wider range of ponds from different localities will fine-tune our findings. The current models should be seen as a first attempt at capturing the factors that determine the productivity of Fingerponds in one quantitative framework. There are several uncertainties about parameter values that limit the use of these models for practical purposes like detailed pond management or policy recommendations. Nevertheless, the models are useful for increasing the understanding of these complex systems and identifying knowledge gaps that need to be filled to achieve further progress in their management. Main areas where the model can be improved are: the importance of nitrogen fixation as an input into Fingerponds; the selectivity and competition for natural food within and between fish species; and the dynamics of detrital pathways.

This study focused on the construction of a dynamic model, finding parameter values for the various rate equations and calibrating the model to the observed values in Fingerponds in Kenya and Uganda. Further work on the model should include sensitivity analysis and validation with independent datasets. If validated, the model can be a useful tool to predict the combined effects of various management measures and environmental conditions on fish production.

Theoretically, it will be possible to simulate scenarios with varying manure inputs, manure quality, stocking densities of fish, sex ratios of tilapia, and harvesting strategies.

Conclusions about organic manure application on fish production

The application of organic manure stimulates natural food production in Fingerponds. It results in an increase in primary productivity and chlorophyll *a*. At the same time, productivity is greatly limited by the high turbidities. The simulation model, though in preliminary stages confirms that with proper fingerpond management fish growth can be greatly enhanced to average fish yields of 1500 to 2800 kg ha^{-1} for a 200 to 300 day culture period. Apart from main differences in hydrological regimes, the main processes within lake and river floodplain ponds can be assumed to be the same. Provided that the model is developed further, it can be used as a management tool for implementation of Fingerponds on a wider scale and can serve as a tool for decision making in their management.

Acknowledgements

This research was funded by the European Union, EU/INCODEV Fingerponds Project No. ICA4-CT-2001-10037 and in part by the Netherlands Government through the Netherlands Fellowship Programme through UNESCO–IHE, Institute for Water Education, Delft, The Netherlands. We would like to acknowledge all Fingerpond partners who contributed to this work.

References

Abdallah, A.A.F., McNabb, C.D. and Batterson, T.R. 1996. Ammonia dynamics in fertilized fish ponds stocked with Nile Tilapia. *The Progressive Fish-Culturist* 58: 117-123.

Acosta-Nassar, M.V., Morell, J.M., Corredor, J.E., 1994. The nitrogen budget of a tropical semi-intensive freshwater fish culture pond. *J. World Aqua. Soc.* 25: 261-270.

Azim, M.E. 2001. *The potential of periphyton-based aquaculture production systems.* Ph.D Thesis, Fish Culture and Fisheries Group, Wageningen Institute of Animal Sciences, Wageningen University, The Netherlands. 219 pp.

Batterson, T.R., McNabb, C.D., Knud-Hansen, C.R., Eidman, H.M. and Sumatadinata, K. 1989. Indonesia: Cycle III of the Global Experiment In *Pond Dynamics/Aquaculture CRSP* Data Report, Vol. 3, No. Egna, H.S.(ed.), Oregon State University, Corvallis, 135 pp.

Boyd, C.E. 1990. *Water quality in ponds for aquaculture.* Birmingham Publishing Co., Birmingham, Alabama. 482 pp.

Boyd, C.E. 1985. Pond evaporation. *Transactions of the American Fisheries Society* 114: 299-303.

Boyd, C.E. 1982. Hydrology of small experimental fish ponds at Auburn, Alabama. *Transactions of the American Fisheries Society* 111: 638-644.

Boyd, C.E., 1972. Sources of CO_2 for nuisance blooms of algae. *Weed Sci.,* 20(5): 492-297.

Brummett, R.E. 2002. Comparison of African tilapia partial harvesting systems. *Aquaculture 214* pp. 103-114.

Brummett, R.E. 1995. Environmental regulation of sexual maturation and reproduction in tilapia. *Rev. Fish. Sci.* 3. pp. 231-248.

Chen, C.W. and Orlob, G.T. 1975. Ecological simulation for aquatic environments. In: B.C. Patten (ed.). *System analysis and simulation in ecology.* III. Academic press, New York, N.Y. pp. 476-587.

Chifamba, P.C. 1990. Preference of *Tilapia rendalli* (Boulenger) for some aquatic plants. *Journal Fish Biology* 36: 701-705.

Delincé, G. 1992. *The ecology of the fish pond ecosystem with special reference to Africa.* Kluwer Academic Publishers, Dordrecht. 230 pp.

Denny, P., Kipkemboi, J., Kaggwa, R. and Lamtane, H. 2006. The potential of fingerpond systems to increase food production from wetlands in Africa. *International Journal of Ecology and Environmental Sciences* 32(1): 41-47.

Edwards, P. 1993. Environmental issues in integrated agriculture-aquaculture and wastewater fed fish culture systems: In: Pullin, R.S.V., Rosenthal, H., Maclean, J.H.L. (Eds.), Environment and aquaculture in developing countries. *ICLARM Conf. Prod.,* Vol 31. ICLARM, Manila, pp. 139-170.

El Samara, M.J. and Oláh, J. 1979. Significance of nitrogen fixation in fish ponds. *Aquaculture* 18: 367-372.

Gangadhar, B., Keshavanath, P., Ramesh, T.J. and Priyadarshini, M. 2004. Digestibility of bamboo grown periphyton by carps and tilapia. *Journal of Applied Aquaculture* 15: 151-162.

Getachew, T. 1987. A study on an herbivorous fish, *Orechromis niloticus* L., diet and its quality in two Ethiopian Rift Valley lakes, Awassa and Zwai. *J. Fish. Biol.* 30: 439-449.

Green, B.W., Phelps, R.P. and Alvarenga, H.R.1989. The effect of manures and chemical fertilizers on the production of *Oreochromis niloticus* in earthen ponds. *Aquaculture* 76: 37-42.

Gross, A., Boyd, C.E. and Wood, C.W. 2000. Nitrogen transformations and balance in channel catfish ponds. *Aquacultural Engineering* 24: 1-14.

Hargreaves, J. A. 1998. Nitrogen biochemistry of aquaculture ponds. *Aquaculture* 166: 181-212.

Jamu, D.M. 1998. Modelling organic matter and nitrogen dynamics in integrated aquaculture/agriculture systems: effects of cycling pathways on nitrogen retention and system productivity. PhD thesis, University of California, Davis, USA.

Jamu, D.M. and Piedrahita, R.H. 2002. An organic matter and nitrogen dynamics model for the ecological analysis of integrated aquaculture/agriculture systems: I. model development and calibration. *Environmental Modelling and Software* 17 (6): 571-582.

Jauncey, K., and Ross, B. 1982. A guide to Tilapia feeds and feeding. University of Sterling, Scotland. 111 pp.

Jiménez-Montealegre, R., Verdegem, M.C.J., van Dam, A., Verreth, J.A.J., 2002. Conceptualization and validation of a dynamic model for the simulation of nitrogen transformations and fluxes in fish ponds. *Ecological Modelling* 147: 123-152.

Jiménez-Montealegre, R.A. 2001. Nitrogen transformations and fluxes in fish ponds: a modeling approach. Ph.D Thesis, Wageningen University, The Netherlands. 185 pp.

Kipkemboi, J. 2006. Fingerponds: Integrated seasonal aquaculture in East African Fresh water wetland. Exploring their potential for wetland wise-use strategies. PhD Thesis, UNESCO-IHE Institute for Water Education, The Netherlands (in prep.).

Knud-Hansen, C.F., Batterson, T.R., McNabb, C.D., Harahat, I.S., Sumantadinata, K. and Eidman, H.M. 1991a. Nitrogen input, primary productivity and fish yield in fertilized freshwater ponds in Indonesia. *Aquaculture* 94: 49-63.

Knud-Hansen, C.F., McNabb, C.D., Batterson, T.R., 1991b. Application of limnology for efficient nutrient utilization in tropical pond aquaculture. *Verh. Internat. Verein. Limnol.* 24: 2541-2543.

Lin, C.K., Tansakul, V. and Apihapath, C. 1988. Biological nitrogen fixation as a source of nitrogen input in fish ponds. pp. 53-58. In: R.S.V. Pullin, T. Bhukaswan, K. Tonguthai and J.L. Maclean (eds.). The Second International Symposium on Tilapia in Aquaculture. *ICLARM* Conference Proceedings 15. Department of Fisheries, Bangkok, Thailand, and International Centre for Living Aquatic Resources Management, Manila, Philippines.

Lorenzen, K., Struve, J. and Cowan, V.J. 1997. Impact of farming intensity and water management on nitrogen dynamics in intensive pond culture: a mathematical model applied to Thai commercial shrimp farms. *Aquaculture Research* 28: 493-507.

Oh, S.Y., Jo, J.Y., Park, J., Noh, K. and Kim, I.B. 2001. Effects of body weight, ration size and dietary protein level on ammonia excretion and faecal production by Nile tilapia, *Oreochromis niloticus*. 6th Asian Fisheries Forum, Book of Abstracts, *Asian Fisheries Society*, Manila. 117 pp.

Pokornỳ, J. Přikryl, I., Faina, R., Kansiime, F., Kaggwa, R.C., Kipkemboi, J., Kitaka, N., Denny, P., Bailey, R., Lamtane, H.A. and Mgaya, Y.D. 2005. Will fish pond management principles from the temperate zone work in tropical Fingerponds. In: J. Vymazal. *Natural and Constructed Wetlands: Nutrients, Metals and Management.* Backhuys Publishers, Leiden, The Netherlands. pp. 382-399.

Poole, H.H. and W.R.G. Atkins, 1929. Photo-electric measurements of submarine illumination throughout the year. *Journal of the Marine Biological Association,* 16: 297-324.

Robotham, P.W. J. 1990. Trophic niche overlap of fry and juveniles of *Oreochromis leucostictus* (Teleostei, Chichlidae) in the littoral zone of a tropical lake (L. Naivasha, Kenya). *Revue Hydrobiologie Tropicale* 23: 209-218.

Schroeder, G.L. 1978 Autotrophic and heterotrophic production of microorganisms in intensely-manured fish ponds, and related fish yields. *Aquaculture* 14: 303-325.

Schroeder, G.L., 1987. Carbon and nitrogen budgets in manured fish ponds on Israel's coastal plain. *Aquaculture* 62: 259-279.

Schroeder, G.L., Alkon, A., Laher, M., 1991. Nutrient flow in pond aquaculture systems. In: Brune, D.E., Tomasso, J.R (eds)., Aquaculture and Water Quality. *World Aquaculture Society*: 489-505.

Schroeder, G.L., Wohlfarth, G., Alkon, A., Helevy, A. and Krueger, H. 1990. The dominance of algal-based food webs in fish ponds receiving chemical fertilizers plus organic manures. *Aquaculture* 86: 219-229.

Snedecor, G.W. & Cochran, W.G. 1980. *Statistical methods*, 7th edition. The Iowa State University Press, Ames, Iowa. 507 pp.

Stickney, R.R. 1996. Tilapia update 1995. *World Aquaculture Society* 27(1): 45-50.

Tadesse, Z. 1999. The nutritional status and digestability of *Oreochromis niloticus* L. diet in Lake Langeno, Ethiopia. *Hydrobiologia* 416: 97-106.

Teichert-Coddington, D.R., Green, B.W., and Phelps, R.P. 1992. Influence of site and season on water quality and tilapia production in Panama and Honduras. *Aquaculture* 105: 297-314.

Trewavas, E.T. 1983. *Tilapiine fishes of the genera Sarotherodon, Oreochromis and Danakilia*. British Museum (Natural History) Publication, London UK. 583 pp.

Tudorancea, C.R., Fernando, C.H. and Paggi, J.C. 1988. Food and feeding ecology of *Oreochromis niloticus* (LINNAEUS, 1758) juveniles in Lake Awassa (Ethiopia). *Arch. Hydrobiol. Suppl.* 79: 267-289.

U.S Soil Conservation Service. 1972. *Engineering Handbook.* Section 4: Hydrology. Washington, D.C. 422 pp.

van Dam, A.A. & M.C.J. Verdegem, 2005. Utilization of periphyton for fish production in ponds: a systems ecology perspective. In: M.E. Azim., M. C. J. Verdegem., A. A. van Dam., M. C. M, Beveridge (eds.) *Periphyton: Ecology, exploitation and management.* CABI publishing. pp. 91-111 pp.

van Dam, A.A., Dardona, A.W.J and Kansiime, F. 2004. Nitrogen retention modeling for papyrus wetlands. *Project Report. Scientific results of the Ecotools Project.* Lake Victoria wetland and inshore areas. 269 pp.

Yoo, K.H., and Boyd, C.E. 1994. *Hydrology and water supply for pond aquaculture.* New York, NY: Chapman and Hall. 483 pp.

Chapter 7

General Discussion and Conclusions

Introduction

Wetlands play an important role in ameliorating food scarcity mainly through crop production and fish culture. If well-exploited, wetlands can contribute significantly to the achievement of some of the Millennium Development Goals (MDGs), particularly the eradication of hunger and extreme poverty. In an attempt to address poor food security facing many people in Sub-Saharan Africa, and East Africa in particular, the potential of wetlands for fish culture was explored using seasonal wetland fish ponds 'Fingerponds'. Fingerponds can be described as agro-piscicultural systems modified from current schemes developed in Mexico, China, Asia and more recently Africa (Denny *et al.*, 2006). They maximize the agriculture and aquaculture potential of natural wetlands. Processes governing the functioning of these ecosystems were assessed in order to devise management strategies suitable for resource-poor communities. This thesis sought to determine how nutrient dynamics and primary productivity regulate fish production in Fingerponds. It evaluated their functioning and dynamics based on the application of natural organic manure (chicken and fermented green biomass) and use of artificial substrates to encourage the growth of periphyton.

This study demonstrated the effectiveness of organic manure inputs and periphyton substrates in enhancing natural food for fish in Fingerponds. Primary productivity (periphyton and phytoplankton) was enhanced with increments of up to 92 %, leading to resultant average net primary productivities of up to 2.9 g C m^{-2} d^{-1} in manured ponds. These values are comparable with other organically manured ponds (Egna & Boyd, 1997). However, high clay turbidity, particularly at one of the locations (Gaba in Kampala), led to light attenuation lowering the effective productive layer in the ponds. The study also showed that with increased nutrient levels and enhanced phytoplankton productivity, fish growth and production were greatly improved.

A number of drawbacks was found in carrying out my field research: these included; an uncontrollable functional period of ponds, high recruitment and stunting of the fish, and poor water quality (e.g. high turbidities, nitrogen limitation). This chapter reviews these issues outlining their implications, analyzes difficulties encountered in the research methods (such as variability and lack of replication) and puts forward recommendations for improvement. The link between aquaculture development and food and nutritional security is underscored and policy issues for wetland management, food security and poverty alleviation outlined. An assessment of the gaps in knowledge is made and future perspectives highlighted. Finally, new areas of research are delineated. A synopsis of the main findings is given prior to the main discussion.

Synopsis of main findings of this thesis

The first chapter considered issues pertaining to food security in Sub-Saharan Africa and highlighted the status of aquaculture in the region. It described the role and potential for fish culture in wetlands as a means to ameliorate food scarcity. The Fingerpond concept was described. It was argued that natural wetlands are vital for fisheries and that integrated agriculture-aquaculture systems are a step in the right direction in tackling the severe food security problem in Sub-Saharan Africa.

The potential of the wetland ecotone for fish culture through seasonal fish ponds 'Fingerponds' was demonstrated in Chapter 2. Key factors determining the conduciveness of Fingerponds for fish culture such as soil conditions, seasonal flooding, natural stocking, and water quality were examined. Fingerponds depend upon the seasonal floods for fish from lakes or rivers to migrate to the ponds. In our study this proved successful, the flooding of the fringing wetlands governed by lake levels plays a major role in determining the fish species stocked. The main ecological processes determining water quality in the freshwater wetland ecotone and Fingerponds were found to be buffering capacity and sedimentation/re-suspension. Finally, the need to enhance nutrient levels and primary productivity in order to sustain good fish growth throughout the dry season was demonstrated.

Use of organic manure (chicken and green manure) to enhance fish production was explored in Chapter 3. Application of chicken manure at rates of 520 to 1,563 kg ha^{-1} every two weeks increased phytoplankton primary productivity without adversely affecting pond water quality; concentrations of oxygen and ammonia remained within limits acceptable for the survival of the fish. The main limitations to phytoplankton primary productivity were high suspended solids/turbidity, decreasing water levels and nitrogen limitation.

In Chapter 4, an alternative flux for nutrients was explored: the 'periphyton loop'. This was achieved using artificial substrates for periphyton attachment; bamboo and three wetland plants (*Phragmites,* papyrus and *Raphia*). Bamboo and *Phragmites* proved to be the superior substrates. Despite high water turbidities, increased substrate surfaces within the ponds resulted in more than doubling of primary productivity (phytoplankton and periphyton) in manured ponds. Periphyton contribution to total primary productivity was as high as 92% of the total primary productivity (0.90 g C m^{-2} d^{-1}) in Walukuba. Periphyton provided an additional, more diverse source of natural food for the fish.

Chapter 5 looked at fish production in organically manured Fingerponds in the presence of the substrates. A maximum yield of 2.67 tonnes ha^{-1} yr^{-1} was obtained. The main limitations to fish growth were: high reproduction/recruitment of fingerlings (due to high feeding pressure and subsequent stunting of fish) resulting in small sized fish; low water levels, high light limitation due to turbidity (hence lowered primary productivity) and low zooplankton biomass. An increased ratio of male to female fish was obtained although manual sexing of fish and periodic removal of female fish and small fish (total lengths < 5 cm) were unsuccessful in curbing reproduction. At the final harvest, per capita fish consumption was enhanced even though the fish did not attain 'table' size (total length of about 15 cm).

A systems modelling approach was adopted in Chapter 6 to give insight into processes that determine fish growth and to relate fish yields to the natural food production and management of Fingerponds. Two types of fingerponds were distinguished by differences in hydrological regimes: lake floodplain fingerponds and those also fed by inflowing rivers referred to as river floodplain fingerponds. The ground water levels vary and for ponds near the lake the hydrology is governed strongly by lake seiches. The daily seiche movements provide water recharge in the wetland and consequently, these ponds do not dry out unless the drop in lake levels is significant. Nitrogen budgets were computed and quantitative estimates of all process flows given. The simulation model, though in preliminary stages confirmed that with proper fingerpond management fish growth can be greatly enhanced to average fish yields of 1,500 to 2,800 kg ha^{-1} for a 200 to 300 day culture period. Apart from main differences in hydrological regimes, the main processes within lake and river floodplain ponds can be assumed to be the same. Provided that the model is developed further, it can be used as a management tool for implementation of Fingerponds on a wider scale and can serve as a tool for decision making in their management.

Major issues pertaining to the functioning and management of Fingerponds and their implications

Functional period of Fingerponds

Fingerponds are solely dependent on natural flood events, thus involving an element of risk. Their functional period is governed by external factors such as climatic or environmental changes that are not easily controllable. Nevertheless, most areas experience normal flood events which are reasonably predictable so this should not present a major set back. However, catastrophic changes can occur, as shown in this study (i.e. the sudden drop in Lake Victoria levels by >1.5 m during the 2003-2005, period; Chapter 5, Figure 1) that shorten the functional period. In addition to floods, precipitation and the ground water hydrology play a major role in determining the functional period of a Fingerpond. It would be helpful to develop simple techniques that a farmer can use to estimate the functional period of Fingerponds based on pond water level measurements. The current design of Fingerponds does not allow for easy filling of ponds during variable flood events. One could consider changing the design, e.g. by creating a gentler slope from the wetland towards the shallow end to allow for easier filling/stocking of ponds when the floods recede, or by use of small inlet channels as was the case in chapter 2. Improvements in Fingerponds design should always leave the natural functions of the wetland intact.

Recruitment and stunting

A risk in the culture of tilapias in fishponds is the high recruitment of fish. As shown in this study, high recruitment rates limit fish growth with final harvests comprising mainly small-sized fish amounting to over 20% of the final harvest. The high feeding pressure arising from these numerous fishes limits the size to which the fish can grow. Additionally, females do not grow big as most of their energy is spent on reproduction, resulting in 'stunting' of fish.

Other studies on curbing recruitment of Nile tilapia using large active predators such as *Clarias gariepinus* (Burchell, 1822) and *Ophiocephalus obscures* have been successful (de Graaf et al., 1996) but the predator-prey ratios have to be high for effective predation (de Graaf et al., 1996; de Graaf & Janssen, 1996). In this study very few active predators, such as *C. gariepinus,* entered the ponds despite their being in the wetlands towards the wetland-lake interface. *Clarias* tend to move along a current and may require good floods to bring them into the ponds. If no *Clarias* stock the ponds, their fingerlings can be collected and transferred from the wetland-lake-interface. Regardless of its predatory nature, *C. gariepinus* can cease to be an active predator if provided with an adequate alternative food source like insects, macrophytes, detritus and zooplankton (Dadebo, 2000; Graaf *et al.*, 2005). With such low zooplankton biomasses in naturally stocked Fingerponds, *Clarias* are likely to shift to a more piscivorous behaviour when still relatively small (< 8 g). *Clarias* have the added advantage that they are also a highly appreciated food by the local people.

Other options for curbing recruitment such as periodic harvesting of fish fry and fingerlings using mosquito nets or traditional methods (e.g., mixing of water with a stick or waddling in the pond) can be employed. For the latter, care must be taken not to cause oxygen depletion and is best done when water levels are relatively high (> 0.5 m). Manual sexing can be adopted for the removal of females but is only effective for bigger fish (>30 g). The farmer has to develop the skill of differentiating between male and female tilapias. However, the local people were able to learn how to sex fish after being shown how to do it using simple pictorial guides (Plate 7.1.). Low-cost alternatives of curbing recruitment of tilapias in Fingerponds need to be further investigated.

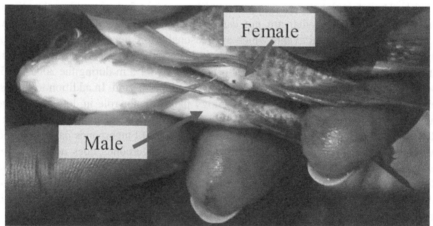

Plate 7.1. Manual sexing of tilapias during one of the fish censuses. Photograph by R.C.Kaggwa

Water quality
Good water quality is crucial for fish production. In this study, high clay turbidity was a major limitation: it increased light attenuation through the water and resulted in a lowered effective primary productivity (phytoplankton and periphyton). The application of organic manure to Fingerponds did not lower turbidity significantly nor did the planting of grass on the banks to reduce run-off effects, but it did reduce soil erosion. Run-off from pond dikes (banks) has been reported to contribute significantly to clay turbidity in fish ponds (Yi *et al.*, 2003). Organic matter (e.g. dead vegetation or barnyard manure) has been used in fishponds to reduce turbidity but it has to be applied carefully as it exerts an oxygen demand (Boyd, 1990). In closed systems such as Fingerponds, application of extra organic matter may trigger oxygen depletion and must be done carefully. This can be achieved by adding the

organic matter at selected intervals based on the colour of the water and the behaviour of the fish. From our findings it is apparent that the clay turbidity in these ponds arose from the soil type. More research is required on low-cost methods of reduction of turbidity that are easily adoptable by farming households.

The concentrations of oxygen and ammonia in the water are critical for fish survival and growth. In Fingerponds, they are best controlled by regulated application of organic manure (animal and green manure). To avoid excessive nutrient enrichment, that reduces oxygen levels and increases ammonia concentrations, it is imperative to reduce the application rates as the growth period progresses and water levels fall. Close monitoring of the behaviour of fish, particularly at dawn, can be a good indicator of the water quality. In addition, fish gills can be checked regularly for signs of ammonia toxicity. Pond transparency can be checked easily (using a secchi disk or a hand palm) and when it is low due to excessive algal growth, manure applications can be reduced.

Research methods and difficulties

Variability, replication and field research

In attempting to ensure sustainable wetland management through integrated aquaculture-agriculture systems, it is imperative that research be done that combines modern science with old traditional methods. The combined knowledge, practices and regulations foster a closer interaction of societies with the environment. Aquaculture ponds are unique systems that respond differently to fertilization even when in the same location and treated identically (Boyd, 1990; Knud-Hansen, 1998). In this study, ponds were variable due to a number of factors; differences in fish species/sizes at initial stocking of the ponds, diversity of fish population, pond hydrology (low pond water levels, differences in seepage and infiltration rates), differences in soil types (texture, nutrient levels) and N:P ratios in the water column. Combined pond-to-pond variability and fish-to-fish variability result in loss of information and may provide wrong statistical tests (Riley & Edwards, 1998). Conversely, the usefulness of statistical methods in planning fish culture research and analysis of experimental data cannot be over-emphasized.

In modern research of aquaculture ponds, replication and controls are included wherever possible for the statistical analyses from which appropriate and accurate recommendations are made. Good experimental designs are employed that take into account replication, randomization and independence (Riley & Edwards, 1998). The optimum number of replicate ponds is based on the degree of precision required in estimating the effects of the treatments applied and the expected levels of variability. However, the number of replicates needed to gain sufficient statistical power can be very large and, in reality, is often unattainable (van Dam, 1990). Replication and randomization of treatments come at a cost (e.g. land, management) particularly when combining natural science, applied ecological research and community participation. Without sufficient replication in this study, it was not possible to compare treatments (i.e. ponds). Additionally, a limitation to the sensitivity of treatments was experienced. Furthermore, it was not possible to gauge the precision of the measurements.

Fingerponds, by their very nature, are based in natural wetlands. To obtain research results that are realistic and applicable to actual farm/field situations, it is essential that the research be done in the field and not in research stations. As a consequence, it was not possible to have many ponds because of the high costs and impracticability. To enable a degree of independence of data, separate treatments were set up in different ponds in this study. However, without replication and randomization, it may be argued that independence of data was not achieved since no direct comparison could be made between treatments. This remains debatable.

With a burgeoning need to carry out research and develop applied technologies relevant to the problems of developing countries, a balance must be found between good science and relevance to the context in which the technology is to be applied, encompassing both ecological and social aspects. Multi-sectoral integrated approaches driven by the needs of the people provide a plausible trade-off. These approaches are realistically sustainable and take into account local environments and conditions. To analyze the data, statistical methods that analyze variation in the data and relate this to management options/factors were adopted. In this study, exploratory methods that identify inter-relationships and structures among variables such as factor analysis were used (Milstein, 1993; Prein et al., 1993). However, hypothesized relationships between variables could not always be tested (e.g. with multiple regression, analysis of variance) due to inadequate replication.

Mechanisms and application of pond inputs
Most chicken manure available for use by resource-poor communities is from chickens that scavenge for their sustenance (free-range chicken) and are not well-fed. This results in a poor-quality manure. Additionally, chicken manure deteriorates when poorly stored; aerobic decomposition occurs, resulting in the release of carbon dioxide and ammonia, which reduces the nitrogen and carbon available for primary productivity (Muck & Steenhuis, 1982). With chicken manure purchased from the local market in this study, the quality of chicken manure was predetermined and we had no control over it.

Direct consumption of manure by fish can occur (Knud-Hansen et al., 1991) but the suspended manure particles impair fish production by diminishing the nutritive quality of filterable manure. Furthermore, the addition of manure for direct consumption may degrade the water quality (Knud-Hansen, 1998). In Fingerponds, chicken manure applied through a cage/crib in the corner of the ponds allowed for a regulated gradual release of nutrients which safeguards the system against a sudden drop in oxygen levels. In the event of lowering oxygen concentrations, the manure can easily be scooped out of the cage. Small fish that tend to congregate around the manure cage/crib to graze the periphyton and bacteria can easily be removed with a small scoop net (Bailey, pers comm.) thus providing a good way of reducing the numerous tiny fish too. This method was not applied since the small fish did not congregate around the cage.

When using fermented green biomass, care must be taken not to over-fertilize the pond as the rich organic mixture creates an increased oxygen demand. It is best applied once a week. Fresh green vegetation (e.g. finely cut cassava leaves, cabbage leaves or cocoa yams) or kitchen wastes may be added but must be applied cautiously, and only when it is consumed readily by the fish.

For research purposes, I installed artificial substrates for the attachment of periphyton in the Fingerponds using a mat matrix. Mats (this thesis) and poles inserted in the pond bottom (Azim et al., 2005) are both suitable but should be easily removable to facilitate pond management (e.g. seining of ponds). Twigs, branches of trees or poles made from plant material, such as bamboo and reeds, are a practical, local alternative and can be used as long as the materials are not toxic. A substrate coverage of 30-70% of the pond surface area is recommended (this thesis). Excessive shading that affects adversely the pond water quality must be avoided. Floating macrophytes such as water lilies (*Nymphaea* spp.), water lettuce/Nile cabbage (*Pistia stratiotes*) and grasses all provide good substrates for periphyton attachment and act as a refuge for fish although they must be prevented from covering the whole pond surface (25% coverage of ponds surface is recommended). At the end of the season they should be dug out to suppress their excessive, invasive growth the following season. Substrates that decompose quickly, such as papyrus or *Raphia*, should be removed from the water before they decompose completely. With falling water levels over a grow-out period, rapid decomposition of substrates can lead to increased nutrient levels triggering algal blooms that cause oxygen depletion and subsequent fish kills.

Monitoring of fish stock and growth
In this study, it was not possible to carry out a complete census of the fish at stocking or frequent assessment of fish growth using the seine method. The effectiveness of this method is dependant on the skill of the net operators and pond water levels. With high water levels at the start of the culture period fish are likely to escape from the net. Estimation of fingerlings is equally difficult as nets with much smaller mesh sizes (mesh size 2-5 cm) are required. The assessment of fish growth is also challenging as the fish take refuge within the periphyton substrate matrix. As most resource-poor communities do not have access to fish nets traditional harvesting methods are best employed such as basket trapping, hook and line, or seining with a reed fence (Brummett, 2002). Extension workers can be used to train the local people in this regard.

Aquaculture development: its linkages to food and nutritional security

The development and wider adoption of aquaculture can be seen as a basis for enhancing food security and other needed welfare (Ahmed & Lorica, 2002). Gupta *et al.* (1999) reported over six-fold increments in fish production following an aquaculture development intervention (i.e. integration of rice farming with aquaculture) in Bangladesh whilst Ahmed & Lorica (2002) reported enhanced income and consumption. From this study it is evident that Fingerponds are more likely to be stocked with smaller-sized fish suggesting an enhancement of consumption rather than income. Smaller fish contribute significantly to the diet and nutritional requirements of the rural poor (Brummett & Noble, 1995; Thilsted *et al.*, 1997; Bouis, 2000). In fact, in the Lake Victoria region, most families now end up with the smaller fish, or the skeletons of bigger fish whose fillet are exported or sold to hotels (UNDP, 2005). With increasing demand for bigger fish for the global market, more and more poor communities are left to feed on small-sized fishes (Delgado *et al.*, 2003). Scaling up of Fingerponds may not result in a dramatic increment of big fish but can contribute to an increase in production of medium-sized fish for the lower income groups. Furthermore, this technology can help community members who do not go fishing in the open waters of the Lake.

Fingerponds can contribute significantly to the Millennium Development Goals (MDGs) (Figure 7.1). They increase the per capita fish consumption and in so doing stabilize food security which can be quantified when incorporated into the day-to-day activities of riparian communities. These integrated agriculture-aquaculture systems also contribute to the 'wise use' of wetlands and encourage more efficient use of available resources within natural wetlands. They enhance people's livelihoods and stimulate the involvement and empowerment of local communities. Fingerponds can also affect traditional gender roles in the communities of the Lake Victoria wetlands. Fishing in the wetlands has been traditionally for men. At the Fingerponds' site in Walukuba, near Jinja, Uganda, a women's group was actively involved in the day-to-day management of the wetlands through bye-laws established within the municipality and undertook the management of the Fingerponds. The women were able to learn about fish culture and could access protein for their families as well as vegetables from the raised bed gardens, Fingerponds thus give women direct access to fish and vegetables, providing them with a much needed source of protein and vitamins.

Policy issues for wetland management, food security and poverty alleviation

The relationships between wetland and fish production are an essential and important part of the on-going debate on wetland regulation and policy; establishing the link between wetlands and poverty, involvement of all stakeholders in decision-making and wetland management; particularly the people that depend on these wetlands for their livelihoods (WI, 2006). These relationships are rather complicated and are not always recognized. National policies need to incorporate strategies of 'wise use' of wetlands to ensure both environmental and social sustainability and development. In Uganda, this is taken care of under the National Wetlands Policy of 1995. Fingerponds provide a means of contributing to this although more research is needed to determine their impact on the environment when developed on a larger scale. Nevertheless, Fingerponds are a step away from donor-driven to self-driven development strategies. They offer a low-cost technology easily adoptable by even the poorest communities. The management of the systems is simple and the outputs in terms of protein availability are good. In the long-term they can result in increased income and prosperity.

Fingerponds combat some of the major difficulties experienced in conventional fish pond systems such as water availability and fish stocking costs. One may argue however, that water availability for Fingerponds presents a risk to the farmer. In normal climatic regimes, floods can be a stable source of water for Fingerponds. The water trapped in these ponds can also be used for watering of vegetable plots on the raised beds. With simple management techniques (Chapters 3, 4 and 5), resource-poor communities can manage this resource effectively.

Technological advances almost universally help the rich more than the poor (Kent, 1976). This arises from the failure of policies to correct social and institutional inadequacies or constraints preventing wider participation of small-scale farmers in new technologies. In scaling up the Fingerpond technology this needs to be taken into consideration.

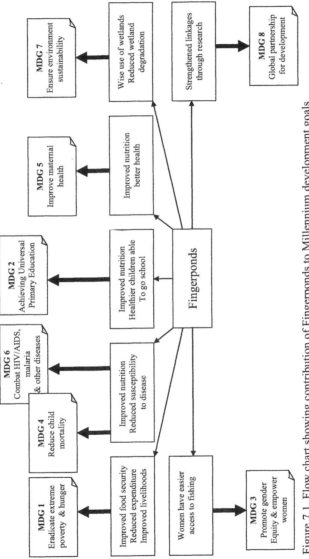

Figure 7.1. Flow chart showing contribution of Fingerponds to Millennium development goals

It can only be addressed through policies that integrate all stakeholders. These policies need to be drawn up fostering wise use of wetlands with a focus on improved fisheries, health, food security and poverty alleviation. It is necessary to take into consideration cross-cutting issues such as land tenure and accessibility, water availability, environment and wetland laws and regulations. Richer people generally establish a greater claim on land and water resources. In many countries ownership of wetlands is not well defined and can present a problem in the transfer of the Fingerpond technology. In Uganda on the other hand, the law stipulates that wetlands "are held in trust" by the Government and Local Governments (GOU, 2000) and the Ugandan National Wetland Policy of 1995 actively encourages the sustainable use of wetlands and fish farming at the edges of wetlands is considered as one of the wise uses. Thus, the scaling up of the Fingerpond technology is to be encouraged with additional research to consider the full social aspects and the possible wider implications for the wetland environment.

Gaps in knowledge and future perspectives

From this study, a number of gaps in knowledge have been identified in regards to the functioning and management of Fingerponds. These provide a springboard for potential research that will deepen the understanding of Fingerponds.

Optimization of fish production based on an integrated use of animal manure, green biomass and periphyton substrates

Taking into account that most poor people may not even own a few chickens, the use of animal manure alone is rarely feasible: it is necessary to integrate the pond systems into the farming and wetland activities. As such, it is strongly recommended that the use of organic manures (animal and green) and periphyton substrates be combined to optimize natural food production in Fingerponds. Further research in these aspects is to be encouraged for the best management strategies of these systems.

Economic implications of organic manure application and artificial substrates use

Fingerponds are a low-cost technology which can be adopted easily by the poorest communities. In using natural food for fish production in these systems, this study has shown that it is best to include some animal manure (e.g. chicken) in combination with artificial substrates to provide food for fish from the growth of periphyton. The question of whether Fingerponds are economically affordable and sustainable has not been addressed in my research. The use of organic manures was based on the assumption that the pond inputs form part of the nutrient resource flows of many households. However, in the absence of readily available animal manure, funds are needed to purchase it. Despite the incredible inventiveness even in the poorest of communities (e.g. through barter trade), an economic cost-benefit analysis at household level is required.

Recruitment control in Fingerponds using active predators

The effectiveness of predation of small fish by *Clarias* (catfish) in closed systems such as Fingerponds is not elucidated. The determination of the predator-prey stocking ratios cognizant of the natural food availability in Fingerponds, as well as the implications of falling water levels in the ponds during the dry season should be assessed.

Application of natural products to improve nitrogen levels in the ponds

Nitrogen limitation in Fingerponds was a major problem that resulted in proliferation of Cyanobacteria (blue green algae) in the ponds. Previous studies have shown that bluegreen algae can be assimilated by tilapias but its toxicity effects are still debatable (Moriarty & Moriarty, 1973; Collins, 1978; Colman & Edwards, 1987; Jewel *et al.*, 2003). High Cyanobacteria biomasses can have detrimental effects on pond water quality (i.e. oxygen depletion) and cause odour and taste problems in fish; or even kill them. To avoid this it is necessary to enhance nitrogen levels in the ponds but animal manure application can result in the enhancement of phosphorus. Addition of natural products such as young shoots of Napier grass (*Pennisetum purpureum* Schmach) can be considered.

Low-cost methods to reduce turbidity in ponds

The research sites in this study were unusual in that they had high turbidity. Clay turbidity arising from colloidal suspensions was a major limitation to primary productivity (phytoplankton & periphyton) in one of the sites. Contrary to expectations, the high turbidities and suspended solids did not reduce significantly with manuring. This is uncommon for fish ponds (Delincé, 1992; Egna & Boyd, 1997). More research can be carried out to determine simple low-cost techniques to reduce such high clay turbidities.

Role of zooplankton in the growth of tilapias in Fingerponds

Zooplankton is said to play a major role in the food web in fish ponds but in the case of Fingerponds, its role was downplayed mainly because the biomasses were low. Zooplankton was rarely found in the fish gut. With numerous small fingerlings the little zooplankton available was quickly consumed resulting in a high turn-over. In order to enhance the transfer of energy from primary to secondary production it may be beneficial to increase zooplankton biomasses in these ponds. Future research is recommended.

Conclusions

This study demonstrated the potential of fringing wetlands for fish culture. Base-line conditions of the ponds and surrounding wetland formed a basis for the formulation of management strategies. It was clear from my study that Fingerponds in wetlands fringing Lake Victoria, Uganda, are liable to one flood per year and are predominantly stocked with *Oreochromis* species. Buffering capacity in the wetland ecotone and sedimentation/re-suspension determined over 65 % of the variability in water quality. Water quality in the ponds was stable and showed little seasonal or spatial variation with parameters critical for fish survival, i.e., pH, ammonia, oxygen and temperature remaining conducive even in the drier parts of the season. Fish yields were enhanced with application of chicken manure (520 to 1,563 kg ha^{-1} (2 weeks)$^{-1}$) in the presence of substrates for periphyton of bamboo, *Phragmites* and *Raphia* (covering 30-70 % of the pond surface area). On average 1,000 kg ha^{-1} of fish was obtained which provides a real benefit in terms of food protein to poor riparian people.

Fish production from Fingerponds can be further enhanced with pond inputs like animal manure, fermented green manure and green vegetation; and with artificial substrates for periphyton growth. It is possible to diversify these inputs using other materials for artificial substrates as long as they have no potential toxicity. It is necessary, however, that factors limiting productivity of the ponds such as high clay turbidity be mitigated. More research is required that addresses these issues in a practical and easily adoptable form for resource-poor communities.

Although the biological and chemical functioning of the Fingerponds are well elucidated in this study, knowledge gaps still remain that need to be explored in order to further augment the understanding of these systems before scaling up of this technology on a wide level. These gaps do not prevent the use of Fingerponds, but it would be beneficial to address them through research. In carrying out research in these new areas, it is important to consider active participation of farmers in various locations, different cultures and in different wetland areas. Emphasis should be laid on integration of modern science and traditional methods in order to enhance such systems that are applicable and adoptable by resource-poor communities.

More insight into the functioning and management of Fingerpond systems can be achieved by modeling the system including all sources of natural food (phytoplankton, periphyton and zooplankton) based on both phosphorous and nitrogen, and also by addressing the knowledge gaps on light limitation of primary productivity and the food selectivity of tilapia and their fingerlings.

References

Ahmed, M. and Lorica, M.H. 2002. Improving developing country food security through aquaculture development-lessons from Asia. *Food Policy* 27: 125-141.

Azim, M.E., Verdegem, M.C.J., van Dam, A.A. and Beveridge, M.C.M. (eds.) 2005. *Periphyton. Ecology, exploitation and management.* CABI Publishing. 319 pp.

Bouis, H. 2000. Commercial vegetable and polyculture fish production in Bangladesh: their impacts on household income and density quality. *Food and Nutrition Bulletin* 21 (4): 482-487.

Boyd, C.E. 1990. *Water quality in ponds for aquaculture.* Birmingham Publishing Co., Birmingham, Alabama. 482 pp.

Brummett, R.E. and Noble, R. 1995. Aquaculture for African smallholders. *ICLARM* Tech. Rep. 46. 69 pp.

Brummett, R.E. 2002. Comparison of African tilapia partial harvesting systems. *Aquaculture* 214: 103-114.

Collins, M. 1978. Algal toxins. *Microbiol. Rev.* 42: 725-746.

Colman, J.A. and Edwards, P. 1987. Feeding pathways and environmental constraints in waste-fed aquaculture: balance and optimization. In: D.J.W. Moriarty & R.S.V. Pullin (eds.) Detritus and microbial ecology in aquaculture. *ICLARM Conference Proceedings* 14, pp. 240-281.

Dadebo, E. 2000. Reproductive biology and feeding habits of the catfish *Clarias gariepinus* (Burchell) (Pisces: Clariidae) in Lake Awassa, Ethopia. *Ethiopian Journal of Science* 23 (2): 231-246.

Denny, P., Kipkemboi, J., Kaggwa, R. and Lamtane, H. 2006. The potential of fingerpond systems to increase food production from wetlands in Africa. *International Journal of Ecology and Environmental Sciences* 32(1): 41-47.

de Graaf, G.L. and Janssen, J.A.L. 1996. Artificial reproduction and pond rearing of the African catfish *Clarias gariepinus* in Sub-Saharan Africa. A handbook. *FAO Fisheries Technical Paper* 362. Rome, *FAO*. 73 pp.

de Graaf, G.J., Galemoni, F. and Banzoussi, B. 1996. Recruitment control of Nile tilapia, *Oreochromis niloticus*, by the African catfish, *Clarias gariepinus* (Burchell 1822), and the African snakehead, *Ophiocephalus obscuris*. I. A biological analysis. *Aquaculture* 146: 85-100.

de Graaf, G.L., Dekker, P.J., Huisman, B. and Verreth, J.A.J. 2005. Simulation of Nile Tilapia (*Oreochromis niloticus niloticus* L.) culture in ponds through individual-based modeling, using a population dynamics approach. *Aquaculture Research* 36: 355-471.

Delgado, C.L., Wada, M., Rosegrant, M.W., Meijer, S. and Ahmed, M. 2003. *Fish to 2020. Supply and demand in changing global markets*. IFPRI and WFC. 226 pp.

Delincé, G. 1992. *The ecology of the fish pond ecosystem with special reference to Africa.* Kluwer Academic Publishers, Dordrecht. 230 pp.

Egna, H.S. and Boyd, C.E. 1997. *Dynamics of pond aquaculture*. CRC Press, Boca Raton/New York. 437 pp.

GOU, 2000. Wetland and the law. Legislation governing the ownership, use and access to wetlands and their resources. Ministry of Water, Lands and Environment. *The National Wetlands Programme*, Uganda. 39 pp.

Gupta, M.V., Mazid, M.A., Islam, Md.S., Rahman, M. and Hussain, M.G. 1999. Integration of aquaculture into the farming systems of the flood-prone ecosystem of Bangladesh: an evaluation of adoption and impact. *ICLARM*. Technical Report. No. 56.

Jewel, M.A.S., Affan, M.A. and Khan, S. 2003. Fish mortality due to cyanobacterial blooms in an aquaculture pond in Bangladesh. *Pakistan Journal of Biological Sciences* 6 (12): 1046-1050.

Kent, G. 1976. Dominance in fishing. *Journal of Peace Research* 13 (1): 35-47.

Knud-Hansen, C.F., McNabb, C.D., Batterson, T.R., Harachat, I.S., Sumatadihata, K. and Eidman, H.M. 1991. Nitrogen input, primary productivity and fish yield in freshwater ponds in Indonesia. *Aquaculture* 94: 49-63.

Knud-Hansen, C.F. 1998. *Pond Fertilization: Ecological approach and practical application.* PD/A CRSP. 125 pp.

Milstein, A. 1993. Factor and canonical analyses: basic concepts, data requirements and recommended procedures. pp. 24-31 In: M. Prein, G. Hulata & D. Pauly (eds.) *Multivariate methods in aquaculture research: case studies of tilapias in experimental and commercial systems. ICLARM* Stud. Rev.20. Manila, Philippines, 221 pp.

Moriarty, C.M. and Moriarty, D.J.W. 1973.Quantitative estimation of the daily ingestion of phytoplankton by *Tilapia nilotica* and *Haplochromis nigripinnis* in Lake George, Uganda. *J. Zool. Lond.* 171: 15-23.

Muck, R.E. and Steenhuis, T.S. 1982. Nitrogen losses from manure storage. *Agricultural Wastes* 4: 41-54.

Prein, M., Hulata, G. and Pauly, D. (eds.) 1993. *Multivariate methods in aquaculture research: case studies of tilapias in experimental and commercial systems. ICLARM* Stud. Rev. 20. Manila, Philippines. 221 pp.

Riley, J. and Edwards, P. 1998. Statistical aspects of aquaculture research: pond variability and pseudo-replication. *Aquaculture Research* 29: 281-288.

Thilsted, S.H., Roos, N. and Hassan, N. 1997. The role of small indigenous fish species in food and nutrition security in Bangladesh. *NAGA-The ICLARM Quarterly (supplement)* July-Dec. pp. 13-15.

UNDP, 2005. Uganda human development report 2005. Linking environment to human development: A deliberate choice. *UNDP*. 92 pp.

van Dam, A.A. 1990. Is ANOVA powerful enough for analyzing replicated pond experiments? *Aquabyte* 3 (3): 3-5.

WI, 2006. Wetlands and poverty reduction project proposal. Wageningen, The Netherlands. *Wetlands International* 35 pp.

Yi, Y., Lin, C.W. and Diana, J.S. 2003. Techniques to mitigate clay turbidity problems in fertilized earthen fish ponds. *Aquacultural Engineering* 27: 39-51.

Summary

Today the world is facing many challenges among them food security with over 800 million people in the world having too little to eat. A third of these are in Sub-Saharan Africa. Food security is therefore a universal goal but sustainable food production is limited by availability of resources (land, water) and human capacity to increase productivity of these resources without depleting/degrading them. Fish accounts for over 20% of all animal protein in the human diet but in Africa, per capita fish consumption is declining. Furthermore, inland fisheries have also declined. With emphasis on fish for export, more and more of the poorer communities have no access to fish. It is important that food security be increased especially in the drier seasons when livelihoods are most at risk.

At present, aquaculture is ranked amongst the fastest growing segments of the world food economy. But in Sub-Saharan Africa, its development has not been good mainly due to the approach taken in research, development and extension. Emphasis in the past has been on the transfer of technology generated on research stations through rigid administrative structures without allowing farmers to take part. Secondly, focus has been on commercialization which has inevitably left out the poorer communities. Most of the aquaculture systems in place are earthen ponds and yet there is still a big knowledge gap on these production units. In other parts of the world, research on aquaculture systems focuses on enhancement of fish production through the application of organic and inorganic manure and supplementary feeds and more recently, the use of artificial substrates that enhance the periphyton loop within fish ponds. A shift from monoculture systems which have a high production cost and a risk to integrated systems has been made. Integrated agriculture-aquaculture systems use low levels of inputs and are considered semi-intensive. They do no rely on heavy feed and fertilizer inputs and operate in synergy with agriculture (crop-livestock-fish integrated).

The role of wetlands in fisheries has been recognized and in many parts of the world especially Asia. As such Asian countries produce 1,000 times more fish than Africa. In Sub-Saharan Africa this potential has not been fully exploited. Furthermore, with a population growth rate of 2.5%, many ecosystems are under pressure form the needs of a growing population. Wetlands are being encroached upon and are being degraded at an escalating rate. They are best managed through rational compromise between ecological conditions and the levels of human utilization i.e. as 'working wetlands'. This approach underpins the concept of 'wise use' and is based on a form of multi-criteria analysis that integrates biophysical and social economic aspects of wetland utilization. This approach has been tested for agriculture but there is a need to test it in integrated agriculture-aquaculture (IAA) systems too.

Fingerponds are earthen ponds dug from the landward edge of wetlands, extending like fingers into the swamp. They exploit two phenomena: (a) the high natural productivity of the swamp/lake interface (which provides the fish stock for the ponds) and (b) the trapping of fish in depressions following flood retreat on floodplains after seasonal rains. The ponds are self stocking and obviate the necessity of purchasing fingerlings which can be prohibitive for the poorest people. Soil dug from the ponds forms raised beds between the fingers and can be used for cultivation of seasonal crops. The rich organic bottom soil removed from the ponds when they dry up or are drained for harvest, acts as manure for the beds. This "wise use" technology retains the functions of the wetland and can go a long way to ameliorate the poor food security in many parts of Africa.

Fingerponds focus on the development of small-scale farming and preserve the wetland environment in which they are situated while maintaining a productive culture system. They can be incorporated into sustainable fisheries management and provide an alternative model of aquaculture research and development. Fingerpond systems are different from most conventional aquaculture systems in that they are entirely dependant on natural flood events. They thus have seasonal functional periods as water levels are maintained either by precipitation or underground water infiltration during the culture period. In order to stimulate fish production in these systems, the use of natural, organic manures was explored. Application of organic manures to fish ponds results in enhanced algal development, and in closed systems like Fingerponds creates the need for close monitoring and control. Good management practices are required to regulate pond inputs and/or conditions. Additonally, there is need for perceived simplicity or low potential risk, particularly when targeting resource-poor communities. These factors formed the basis for this research.

The main aim of this thesis was to examine the importance of organic manure applications in enhancing nutrient levels, phytoplankton and periphyton productivities and ultimately fish production in seasonal wetland fish ponds, 'Fingerponds'. This was undertaken over a three year period of field and laboratory experiments as outlined in the seven chapters of this dissertation.

This thesis starts with a general introduction (Chapter 1) that provides insight into the current challenges Africa is facing with respect to provision of food security. It highlights the linkage between natural wetlands and fisheries. A definition if Fingerponds is given and the characterization of the study wetlands described. Finally the design and construction criteria for a Fingerpond are outlined.

An assessment was made of the wetland ecotone in order to establish the conduciveness of the environment and its potential for fish culture in Chapter 2. This study investigated the base-line conditions of the ponds and surrounding wetland prior to the formulation of management strategies. Eight 192 m^2 ponds were predominately stocked with *Oreochromis* spp. during the May/June 2003 rains. The flood levels from Lake Victoria played a major role in determining the fish species stocked. The main ecological processes determining water quality in the freshwater wetland ecotone and Fingerponds were buffering capacity and sedimentation/re-suspension which explained over 30% of the variability in water quality. The water quality in the newly established ponds was stable and showed little seasonal or spatial variation. Low nutrient concentrations in the ponds resulted in low phytoplankton biomass. Zooplankton colonization of the ponds was poor resulting in a low biomass. Initial fish stock biomass at the onset of the 2003 grow-out period ranged between 93 and 97 kg ha^{-1}. In order to sustain good fish growth the need to enhance nutrient levels and primary productivity was demonstrated.

Use of organic manure (chicken and fermented green manure) to enhance fish production was explored in Chapter 3. Application of chicken manure at rates of 520 to 1563 kg ha^{-1} every two weeks increased phytoplankton primary productivity without adversely affecting pond water quality; concentrations of oxygen and ammonia remained within limits acceptable for the survival of the fish. Net phytoplankton primary productivity was significantly higher (P<0.05) in manured ponds compared to the unmanured ones and ranged from less than 0.01 to 28.3 g O_2

m^{-2} d^{-1}. Gross phytoplankton primary productivity was 1.1-2.4 fold the net primary productivity. The main limitations to phytoplankton primary productivity were inorganic clay turbidity, decreasing water levels and nitrogen limitation. Low N: P ratios found in the ponds favour dominance of Cyanobacteria which presents a management problem. As the culture period progresses, water levels in the ponds decrease. Thick blooms of blue green algae are likely to cause oxygen depletion and subsequent fish kills. Regulation of manure application rates is vital in keeping this problem in check. This study demonstrated that despite decreasing water levels over the culture period, the pond water quality did not deteriorate to levels unsuitable for fish survival but it might have hampered fish growth.

In Chapter 4, an alternative flux for nutrients was explored through the 'periphyton loop'. This was achieved using artificial substrates for periphyton attachment; bamboo and three wetland plants (*Phragmites, papyrus* and *Raphia*). Bamboo and *Phragmites* proved to be the superior substrates while papyrus was found unsuitable due to its rapid decomposition. The study demonstrated that the highest variability in water quality (18%) arose from the effect of pond inputs/conditions and enhancement of phytoplankton primary productivity in favourable environmental conditions (such as temperature and pond water levels). Furthermore, high inorganic clay turbidity reduced the productivity of the periphyton due to light limitation. Despite these high water turbidities, increased substrate surfaces within the ponds resulted in a doubling of primary productivity (phytoplankton and periphyton) in manured ponds. Periphyton contribution to total primary productivity was as high as 92% of the total primary productivity (0.90 g C m^{-2} d^{-1}) in *Phragmites* based ponds in Walukuba. The study further showed that the presence of periphyton substrates in the ponds did not significantly affect the water quality and that periphyton provided a more diverse source of natural food for the fish.

The potential of periphyton-based organically manured ponds was examined in Chapter 5. Maximum yields of 2.67 tonnes ha^{-1} yr^{-1} were obtained. The study demonstrated that the main limitations to fish growth in Fingerponds were: high reproduction/recruitment of fingerlings (due to high feeding pressure and subsequent stunting of fish) resulting in small sized fish; low water levels, high light limitation due to turbidity (hence lowered primary productivity) and low zooplankton biomass (possibly reducing effective energy transfer). An increased ratio of male to female fish was obtained although manual sexing of fish and periodic removal of female fish and small fish (total lengths < 5 cm) were unsuccessful in curbing reproduction. At the final harvest, per capita fish consumption was enhanced even though the fish did not attain 'table' size (total length of about 15 cm).

A dyamic model was developed in Chapter 6 to give insight into the processes that determine fish growth. The model was simulated for two types of Fingerponds distinguished by the differences in their hydrological regimes; fingerponds also fed by inflowing rivers referred to as river floodplain Fingerponds (viz Kusa and Nyangera, Kenya) and typical lake floodplain ponds (viz Gaba and Walukuba, Uganda). Ground water levels vary and for ponds near the lake the hydrology is governed strongly by lake seiches. The daily seiche movements provide a water recharge in the wetland and consequently, these ponds do not dry out unless the drop in lake levels is significant. The simulation model, though in preliminary stages confirmed that with proper fingerpond management fish growth can be greatly enhanced. Apart from main differences in hydrological regimes, the main processes within lake and river floodplain ponds can be assumed to be the same. The model is preliminary and once subjected to sensitivity analysis and validation can be used as a management tool.

In chapter 7, the results are discussed in light of the original research objectives and implications of Fingerpond management. Drawbacks found in the field research are reviewed and their implications outlined, difficulties encountered in the research methods analyzed and recommendations for improvement made. The link between aquaculture development and food and nutritional security is underscored and policy issues for wetland management, food security and poverty alleviation are outlined. An assessment of the gaps in knowledge is made and future perspectives highlighted. New areas of research are delineated. The main conclusions are: (1) fringing wetlands around Lake Victoria, East Africa have a potential for fish culture and can be further exploited through integrated agriculture-aquaculture systems; (2) fish yields in Fingerponds can be enhanced to 1,500 to 2,800 kg ha^{-1} over a functional period of 200 to 300 days in using animal manure (e.g. chicken), fermented green manure and artificial substrates for periphyton development; (3) clay turbidity, nitrogen limitation and high recruitment are the main factors hindering fish production in Fingerpond systems and once mitigated would result in improved fish yields; (4) knowledge gaps pertaining to light limitation of primary productivity and food selectivity of tilapia and their fingerlings remain (5) the model in its current state is a research tool that identifies knowledge gaps and can be applied to frame hypotheses for further applied research. Ultimately, the model can contribute to improving the management of Fingerponds (6) fingerponds demonstrate the importance of wetlands in providing a livelihood for the ripariarn communities hence the need to conserve wetlands.

Samenvatting

Voedselzekerheid is één van de vele uitdagingen waarmee de wereld momenteel worstelt, met meer dan 800 miljoen mensen die niet genoeg te eten hebben. Een derde van deze mensen woont in Afrika ten zuiden van de Sahara. Voedselzekerheid is een universele doelstelling maar duurzame voedselproductie wordt beperkt door de beschikbaarheid van hulpbronnen (land, water) en het vermogen van mensen om de productiviteit van die hulpbronnen te verhogen zonder ze uit te putten of te beschadigen. Vis vertegenwoordigt 20% van het dierlijke eiwit in het voedselpakket, maar in Afrika loopt de visconsumptie per hoofd van de bevolking terug. De binnenvisserij is ook afgenomen. Nu er meer nadruk ligt op de export van vis hebben steeds meer arme mensen geen toegang meer tot vis. Het is dus belangrijk dat de voedselzekerheid vergroot wordt, vooral tijdens het droge seizoen als het levensonderhoud het meest in gevaar komt.

Aquacultuur, hoewel op dit moment één van de snelst groeiende sectoren van de wereldeconomie, heeft zich in Afrika ten zuiden van de Sahara niet goed ontwikkeld. Dit komt vooral door de benadering die gekozen is in onderzoek, ontwikkeling en voorlichting. In het verleden lag veel nadruk op het overdragen van technologie die ontwikkeld was op onderzoeksstations zonder de participatie van boeren. Verder was er veel aandacht voor commerciële teelten waarbij onvermijdelijk de armere doelgroepen buitengesloten werden. Het meest gebruikte productiesysteem bestaat uit aarden vijvers waarvoor nog steeds een gebrek aan kennis bestaat. In andere delen van de wereld richt het aquacultuuronderzoek zich op verhoging van de visproductie door middel van organische en anorganische meststoffen en bijvoedering, en meer recent ook op het gebruik van kunstmatige substraten die de groei van perifyton in visvijvers bevorderen. Een verschuiving heeft plaatsgevonden van monocultuursystemen met hoge productiekosten en risico naar geïntegreerde landbouw-visteelt systemen, die lagere hoeveelheden mest en voeder gebruiken en als semi-intensief beschouwd worden. Deze systemen hebben geen hoge mest- en voederdoses nodig en functioneren in synergie met landbouwactiviteiten (geïntegreerde vis-, dierlijke en plantaardige productie).

Over de hele wereld, en vooral in Azië, is het belang van wetlands voor de visserij bekend. Azië produceert duizend maal zoveel vis als Afrika. In Afrika ten zuiden van de Sahara wordt het potentiëel voor visproductie niet ten volle benut. Veel ecosystemen staan onder druk door de voedelbehoeften van de populatie die groeit met 2,5 procent per jaar. Wetlands worden in toenemende mate omgezet in landbouwgrond en beschadigd. Wetlands zouden eigenlijk beheerd moeten worden volgens het principe van de `working wetlands´ waarbij een rationeel evenwicht wordt gezocht tussen ecologie en menselijk gebruik. Dit past in een ´wise use´ benadering en is gebaseerd op een multi-criteria analyse die biofysische en sociaal-economische aspecten van wetlandgebruik integreert. Deze benadering is beproefd voor landbouwsystemen maar moet ook in geïntegreerde landbouw-visteelt getest worden.

Fingerponds zijn aarden vijvers die gegraven worden vanaf de landzijde van een wetland en als vingers in het moeras steken. Ze benutten twee verschijnselen: (a) de hoge natuurlijke productiviteit van het overgangsgebied tussen moeras en meer, dat de vis voor de vijvers levert; en (b) het vangen van vis in ondiepe poelen in het moeras na het terugtrekken van de overstroming in het regenseizoen. De vijvers bezetten zichzelf met water en vis en het is niet nodig dure pootvis te kopen. Met de grond die vrijkomt bij het graven van de vijvers worden teeltbedden gevormd voor de productie van groenten. De rijke organische grond, die vrijkomt als de vijvers opdrogen of afgelaten worden voor de oogst, wordt gebruikt als mest voor de groententeelt. Deze 'wise use' technologie behoudt de natuurlijke functies van het wetland en kan een bijdrage leveren aan het verbeteren van de slechte voedselzekerheid in grote delen van Afrika.

Fingerponds richten zich op de ontwikkeling van kleinschalige voedselproductie en tasten het wetlandmilieu niet aan terwijl toch een productief teeltsysteem wordt onderhouden. Fingerponds kunnen onderdeel zijn van een duurzaam visproductiebeheer en vertegenwoordigen een andere benadering van aquacultuur onderzoek en ontwikkeling. Ze verschillen van meer conventionele visteeltsystemen omdat ze volledig afhankelijk zijn van natuurlijke overstromingen. Ze hebben daardoor een seizoensgebonden functionele periode omdat het waterniveau in de vijvers op peil wordt gehouden door neerslag of grondwaterkwel tijdens de teeltperiode. Om de visproductie in deze systemen te stimuleren werd het gebruik van natuurlijke, organische meststoffen onderzocht. Toediening van organische mest aan visvijvers leidt tot de ontwikkeling van algen, en dient in gesloten systemen zoals Fingerponds nauwgezet gecontroleerd en beheerst te worden. Goede beheersmaatregelen zijn nodig om de hoeveelheid meststoffen en de teeltomstandigheden te reguleren. Bovendien is het nodig om de technologie simpel te houden en hoge risico's te vermijden omdat de doelgroep bestaat uit de armere bevolkingsgroepen. Deze overwegingen lagen ten grondslag aan dit onderzoek.

Het belangrijkste doel van dit proefschrift was om het belang te onderzoeken van de toediening van organische mest voor het verhogen van de concentraties van nutriënten, fytoplankton en perifyton en tenslotte van de visproductie in seizoensgebonden visvijvers in wetlands, 'Fingerponds'. Het onderzoek bestond uit veld- en laboratoriumproeven en besloeg een periode van drie jaar zoals beschreven in de zeven hoofdstukken van deze dissertatie.

Het proefschrift begint met een algemene introductie (hoofdstuk 1) die inzicht verschaft in de uitdagingen waarvoor Afrika momenteel staat met betrekking tot het bereiken van voedselzekerheid. De relatie tussen natuurlijke wetlands en visserij wordt belicht. Ook wordt het begrip Fingerponds gedefiniëerd en worden de wetlands die in het onderzoek voorkomen beschreven. Tenslotte worden de vereisten voor ontwerp en constructie van een Fingerpond beschreven.

Om de geschiktheid van het wetlandmilieu voor visteelt te beoordelen wordt in hoofdstuk 2 het wetland als overgangszone tussen meer en land beschreven. De basisomstandigheden voor de vijvers en de wetlandomgeving werden onderzocht voordat een begin werd gemaakt met het formuleren van een beheersstrategie. Acht vijvers, elk 192 m^2 groot, werden bezet met voornamelijk *Oreochromis* spp. gedurende de regentijd van mei-juni 2003. Het overstromingsniveau van het Victoriameer speelde een belangrijke rol in de soortsamenstelling van de vispopulatie. De belangrijkste ecologische processen die de waterkwaliteit in de wetlandzone bepaalden waren de buffercapaciteit en sedimentatie/resuspensie, die 30% van de variatie in waterkwaliteit bepaalden. De waterkwaliteit in de nieuwe visvijvers was stabiel en vertoonde weinig seizoensgebonden of ruimtelijke variatie. Door de lage nutriëntenconcentraties in de vijvers was de fytoplanktonbiomassa laag. Kolonisatie van de vijvers door zoöplankton was slecht waardoor de zoöplanktonbiomassa laag was. De visbiomassa bij de aanvang van de groeiperiode in 2003 was tussen 93 en 97 kg ha^{-1}. De noodzaak om de nutriëntenconcentraties en primaire productiviteit te verhogen teneinde een goede visgroei te bereiken werd aangetoond.

Het gebruik van organische bemesting (kippenmest en compost) voor het verhogen van de visproductie werd onderzocht in hoofdstuk 3. Gebruik van kippenmest in doseringen van 520 to 1563 kg ha^{-1} per twee weken verhoogde de primarie productiviteit van fytoplankton zonder dat de waterkwaliteit verslechterde en de concentraties van zuurstof en ammonia bleven binnen grenswaarden die acceptabel zijn voor de overleving van vis. De netto primaire productiviteit van fytoplankton was significant hoger (P < 0.05) in bemeste vijvers dan in onbemeste vijvers en variëerde van minder dan 0.01 to 28.3 g O_2 m^{-2} d^{-1}. De bruto primaire productiviteit van fytoplankton was 1.1-2.4 maal zo hoog als de netto primaire productiviteit. De belangrijkste beperkingen voor de primarire productiviteit van het fytoplankton werden gevormd door turbiditeit veroorzaakt door anorganische kleideeltjes, lage waterstanden en stikstofbeperking. De lage verhouding tussen stikstof en fosfor die in de vijvers werd aangetroffen bevorderde de dominantie van Cyanobacteria. Dit leidt tot een beheersprobleem, aangezien het waterniveau gedurende de teeltperiode daalt. Extreme bloei van de blauwgroene algen veroorzaakt dan gemakkelijk een gebrek aan zuurstof, en vervolgens vissterfte. Het reguleren van de hoeveelheid mest is essentiëel om dit probleem onder controle te houden. Het onderzoek liet zien dat ondanks het dalende waterniveau gedurende de teeltperiode de waterkwaliteit in de vijvers niet verslechterde tot niveaus die ongeschikt waren voor de overleving van de vis. Wel was er een risico van beperking van de visgroei.

In hoofdstuk 4 werd een alternatieve nutriëntenstroom in de vijvers door middel van perifyton onderzocht. Dit werd gedaan door het gebruik van kunstmatige substraten waarop zich perifyton kan vastzetten. De substraten werden gemaakt van bamboe en drie materialen afkomstig van wetlandplanten (*Phragmites*, papyrus en *Raphia*). Bamboe en Phragmites bleken de beste substraten te zijn terwijl papyrus ongeschikt bleek vanwege de snelle afbraak. Het onderzoek liet zien dat de grootste variabiliteit in waterkwaliteit (18%) werd veroorzaakt door vijverinputs en -beheer en stimulering van de primaire productiviteit van het fytoplankton onder gunstige omstandigheden (temperatuur, waterniveau in de vijvers). Hoge anorganische turbiditeit verminderde de productiviteit van het perifyton door middel van lichtbeperking. Ondanks de hoge turbiditeit resulteerde het verhoogde oppervak in de vijvers in een verdubbeling van de primaire productiviteit (fytoplankton en perifyton) in bemeste vijvers. De bijdrage van perifyton aan de totale primaire productiviteit was 92% (0.92 g C m^{-2} d^{-1}) in vijvers met *Phragmites* substraten in Walukuba. De studie liet ook zien dat de aanwezigheid van substraten in de vijvers geen significante invloed had op de waterkwaliteit en dat peripfyton een gevariëerde bron van natuurlijk visvoer was.

In hoofdstuk 5 werden de mogelijkheden van vijvers met perifyton-substraten en organische bemesting onderzocht. Een maximale visopbrengst van 2.67 ton per hectare per jaar werd behaald. De studie liet de volgende belangrijke beperkingen aan de visgroei in Fingerponds zien: hoge voortplantingssnelheid en recrutering van pootvis leidden, door middel van hoge voedseldruk en vervolgens onderdrukte groei, tot kleine visjes, laag waterniveau, sterke lichtbeperking door turbiditeit (en dus lage primaire productiviteit) en een lage zoöplanktonbiomassa die mogelijk de energieoverdracht naar hogere trofische niveaus beperkt. Een hogere verhouding mannetjes:vrouwtjes werd bereikt maar het met de hand sexen van vis en periodieke verwijdering uit de vijver van vrouwelijke vissen en kleine vissen (met een totale lengte van minder dan 5 cm) leidden niet tot beperking van de voortplanting. De uiteindelijke visoogst droeg bij tot verhoging van de visconsumptie ook al bereikte de vis geen consumptiemaat (d.w.z., een totale lengte van ongeveer 15 cm).

In hoofdstuk 6 werd een dynamisch model ontwikkeld om inzicht te geven in de processen die de visgroei bepalen. Het model werd toegepast op twee typen Fingerponds met verschillende hydrologische regimes: vijvers die gevoed worden door rivieren, zgn. 'river floodplain Fingerponds' (zoals in Kusa en Nyangera, Kenia) en vijvers in het stroomgebied van meren ('lake floodplain Fingerponds', zoals in in Gaba en Walukuba, Oeganda). De vijvers hebben verschillende grondwaterstanden en de vijvers in de buurt van het meer worden beïnvloed door seiches vanuit het meer. De dagelijkse golfbewegingen zorgen voor aanvulling van het grondwater in het wetland, waardoor deze vijvers niet uitdrogen tenzij het waterniveau in het meer daalt. Hoewel het nog verder ontwikkeld moet worden bevestigde het simulatiemodel dat met goede beheersmaatregelen de visgroei kan worden gestimuleerd. Afgezien van verschillen in hydrologische regimes kunnen de belangrijkste processen in de twee typen vijvers verondersteld worden gelijk te zijn. Het model is nog in ontwikkeling en kan, nadat een gevoeligheidsanalyse en validatie zijn uitgevoerd, gebruikt worden als een hulpmiddel bij het beheer.

In hoofdstuk 7 worden de resultaten besproken in het licht van de oorspronkelijke doelstellingen van het onderzoek en worden gevolgtrekkingen gemaakt voor het beheer van Fingerponds. Een overzicht van de beperkingen van het veldonderzoek en de gevolgen daarvan wordt gegeven, de problemen met de onderzoeksmethode worden geanalyseerd en aanbevelingen voor verbetering worden gedaan. Het verband tussen ontwikkeling van de aquacultuur en voedselzekerheid wordt onderstreept, en een overzicht van beleidsaspecten voor wetlandbeheer, voedselzekerheid en armoedebestrijding wordt gegeven. Hiaten in de kennis en verwachtingen voor toekomstige ontwikkelingen worden aangegeven. Ook worden nieuwe onderzoeksgebieden geïdentificeerd. De belangrijkste conclusies zijn: (1) de wetlands rond het Victoriameer in Oost Afrika bieden goede mogelijkheden voor aquacultuur en kunnen benut worden door middel van geïntegreerde systemen voor landbouw en visteelt; (2) de visopbrengsten in Fingerponds kunnen gestimuleerd worden tot 1500-2800 kg per hectare gedurende een periode van 200 tot 300 dagen met toepassing van dierlijke mest (bijvoorbeeld van kippen) of plantaardige compost en kunstmatige substraten ten behoeve van perifyton; (3) de belangrijkste factoren die de visproductie beperken zijn turbiditeit van kleideeltjes in het water, stikstofbeperking en de hoge recrutering van jonge vis. De visproductie zou nog verder kunnen stijgen als deze beperkingen kunnen worden opgeheven; (4) hiaten in kennis blijven bestaan, zoals bijvoorbeeld met betrekking tot de lichtlimitering van de primaire productie en de voedselselectiviteit van tilapia's en hun juvenielen; (5) het ontwikkelde simulatiemodel is een onderzoekshulpmiddel dat hiaten in onze kennis identificeert en gebruikt kan worden om meer toegepast onderzoek te sturen. Uiteindelijk kan het model bijdragen aan verbetering van het beheer van Fingerponds; (6) Fingerponds benadrukken het belang van de wetlands voor de bestaanszekerheid van mensen die in en rond wetlands leven en de noodzaak die wetlands te beschermen en behouden.

Acknowledgements

This research was part of the European Union International Development Corporation Fingerponds project (Contract Number: ICA4-CT-2001-10037) that investigated the wise use of integrated agriculture-aquaculture systems 'Fingerponds' at the edge of East African swamps. Innovative semi-intensive fish production techniques were used to enhance the livelihoods of wetland communities. The project was carried out with six partnering institutions; Egerton University (Kenya), Makerere University Institute of Environment and Natural Resources (Uganda), Dar es Salaam University (Tanzania), ENKI o.p.s (Czech Republic), Kings College, London (United Kingdom) and UNESCO-IHE Institute for Water Education (The Netherlands). The research presented in this thesis was made possible through funding by the European Union, INCO DEV and the Netherlands Government through the Netherlands Fellowship Programme (NUFFIC).

I take this opportunity to thank my promoter, Prof. dr. Patrick Denny for his guidance, able supervision and critical reading of all manuscripts. The long discussions with him over the internet 'SKYPE' were invaluable. I express my esteemed gratitude to Dr. Anne van Dam for his able guidance especially in statistical analysis and ecological modeling. He spent hours working with me both in Uganda and in the Netherlands. Without him a lot of this work would not have been possible. I thank you for your encouragement and jokes which lifted my spirits many a time. I also thank your family for their great hospitality. I thank Dr. Frank Kansiime for according me the opportunity to undertake this research and for his guidance over the years. I take this opportunity to thank Dr. Johan van de Koppel, formerly with UNESCO-IHE for the time and guidance during the proposal writing and Dr. Hans van Bruggen for stepping in his shoes even if for a short time.

I express my sincere gratitude to Dr. William Muhairwe, Managing Director, National Water and Sewerage Corporation (NWSC), Uganda, for allowing me to undertake this PhD under the sandwich programme. I thank the management and staff of NWSC for all the support and facilitation. I especially thank the staff of Kampala-Gaba and Jinja Areas for all technical and moral support during field work, the staff of the Quality Control Department, my supervisors and subordinates whose patience and tolerance enabled me to finish the research. I am grateful to the General Manager, Kampala Water, Mr. Charles Odonga and Dr. Tom Okurut, formally of NWSC for encouraging me to take up this study and Andrew Sekayizzi, Area Manager Jinja for his expert guidance in construction and maintenance of the Fingerponds.

I thank Dr. John Balirwa, the Director Fisheries Resources Research Institute (FIRRI), Uganda, and his staff for all the advice and visits to the sites and Dr. Rose Mugidde (formerly with FIRRI) for guidance on primary productivity. Research in tropical wetlands can in many instances present practical challenges that can only be tackled with dedication and commitment. I am indebted to the able-bodied men that worked tirelessly to dig the ponds, the field assistants Ismail and Robert for their tremendous drive and spirit, Gregory and Dennis who joined the bandwagon many a time with no incentive, Ambrose Wamadere (RIP), Mukwaya and Odeke who spent endless hours behind the wheel and in the field. I thank Mugume for stepping in as a driver whenever I needed one, Katimbo, Arwata, Hawa and Stephen for all the work and endless errands they had to run. I thank the M.Sc students, Austin Mthetiwa,

Grace Sanyu and Deborah Kasule for their contribution to this work. I also thank Anne Serunjoji, for her contribution to the hydrology studies.

I acknowledge all the help accorded to me by the staff of UNESO-IHE who made my travel and stay in Delft comfortable: Jolanda Boots, Fatimah, Jaap, Harry and Dennis. I also thank my fellow PhD students with whom we spent happy and sad moments: Julius Kipkemboi, Richard Buamah, Margaret Abira and Sonko Kiwanuka.

Finally I thank my parents, sisters (Lorna and Barbara), my brother Richard, and brother-in-law Andrew, for their inspiration, love, encouragement and financial support over the years.

Rose Christine Kaggwa
June 2006

About the author

Rose Christine Kaggwa was born on 3rd January 1969 in Tororo District, Uganda. She finished her ordinary level education at Gayaza High School in 1985 and her advanced level education in 1988 at the same school. In October, 1988 she joined Makerere University Uganda where she was awarded a Bachelor of Science Degree in Industrial Chemistry. She was the only female in this class of fourteen. From October 1991 to December, 1991 she worked as a trainee at Pepsicola Ltd and from January, 1992 to April, 1992 as a sales representative with Primex Ltd and Kungu Farm. In May, 1992 she joined National Water and Sewerage Corporation (NWSC) as a Chemist/Analyst and in October 1992 was appointed Plant Chemist of Gaba II waterworks. She was sent for a six-week training course in the United Kingdom where she underwent training at the Water Training International institutions in Derby and at Tadley Court. In 1995 she was transferred to the Water Quality Section of NWSC as a Senior Chemist (Operations). In 1996 she received a scholarship from the Netherlands Fellowship Programme (NFP) to follow an MSc Programme at the International Institute for Infrastructural Hydraulic and Environmental Engineering (IHE), Delft, The Netherlands. In April 1997 she became one of the four pioneer students from IHE to attend the joint programme between IHE and the Austrian Academy of Sciences, Institute of Limnology Mondsee. Her MSc research focused on "The Effect of Water Treatment Sludge on a *Cyperus papyrus*, L. swamp, Kampala, Uganda.' In which she graduated with a Master of Science Degree with distinction in June 1998. She returned to NWSC and worked as a Senior Chemist. In November, 2001 she begun her PhD research program at IHE Delft in collaboration with Wageningen University financially supported by the European Union (EU) INCO-DEV Fingerponds project and the Netherlands Government. In December 2003 she was promoted to Principal Analyst. In January 2004 she was appointed to head the Water Production and Quality Department in Kampala. She held this post until May 2005 when she proceeded on study leave to UNESCO-IHE for her thesis write up. She also held the position of Task Leader for the Industrial and Municipal Waste Management Sub-component of the Lake Victoria Environmental Management Project (LVEMP) for Uganda from 1998 to 2004. Rose is currently heading the External Services Unit of NWSC.

Publications

Kaggwa, R.C., Kasule, D., van Dam, A.A. and Kansiime, F. 2006. An initial assessment of the use of wetland plants as substrates for periphyton production in seasonal wetland fishponds in Uganda. *International Journal of Ecology and Environmental Sciences* 32(1): 63-74.

Kaggwa, R.C., Kansiime, F., Denny, P. and van Dam, A.A. 2005. A preliminary assessment of the aquaculture potential of two wetlands located in the northern shores of Lake Victoria, Uganda. In: J. Vymazal (eds.) *Natural and Constructed Wetlands: Nutrients, Metals and Management*. Backhuys Publishers, Leiden, The Netherlands, pp. 350-368.

Kaggwa R. C., Mulalelo C.I., Denny P., T.O Okurut. 2001. The impact of Alum discharges on a natural tropical wetland in Kampala. *Water Research* 35(3): 795-807.

Other publications

Denny, P., Kipkemboi, J., **Kaggwa, R**. and Lamtane, H. 2006. The potential of fingerpond systems to increase food production from wetlands in Africa. *International Journal of Ecology and Environmental Sciences* 32(1). pp. 41-47.

van Dam, A.A., **R.C. Kaggwa** and J. Kipkemboi. 2006. Integrated pond aquaculture in Lake Victoria wetlands. In: M. Halwart and A.A. van Dam (eds.) 2006. *Integrated irrigation and aquaculture in West Africa. Concepts, practices and potential*, Rome, FAO. 196 pp.

Bailey, R., **Kaggwa, R.**, Kipkemboi, J. and Lamtane, H. 2005. Fingerponds: An agrofish polyculture experiment in East Africa. *AquaNews 32.* University of Stirling. Institute of Aquaculture. pp. 9-10.

Kaggwa, R.., Kiwanuka, S., Okurut, T.O., Bagambe, F., and Kanyesigye, C. 2004. Experiences in setting up ecosan toilets in shoreline settlements in Uganda. In: Werner, C., Avendaño, S. Demsat., I. Eicher., L. Hernandez., C. Jung., S.Kraus., I. Lacayo., K. Neupane., A. Rabiega and M. Wafler. (eds.). Ecosan-closing the loop. Proceedings of the 2nd international symposium on ecological sanitation, incorporating the 1st IWA specialist group conference on sustainable sanitation, 7th-11th April, 2003, Lübeck, Germany. Deutsche Gesellschaft für Technische Zusammenarbeit (GTZ) GmbH, Gemany. pp. 309-316.

Pokorný, J. Přikryl, I., Faina, R., Kansiime, F., Kaggwa, R.C., Kipkemboi, J., Kitaka, N., Denny, P., Bailey, R., Lamtane, H.A. and Mgaya, Y.D. 2005. Will fish pond management principles from the temperate zone work in tropical Fingerponds. In: J. Vymazal. *Natural and Constructed Wetlands: Nutrients, Metals and Management*. Backhuys Publishers, Leiden, The Netherlands. pp. 382-399.

Okurut, T.O., Marron, F., **Kaggwa, C.R.**, and Okwerede, E.L. 2003. Determination of industrial and municipal effluent and urban run-off loads discharged to Lake Victoria from the Ugandan catchment. *The African Journal of Tropical Hydrobiology and Fisheries.* 2. special issues. pp 54-66.

Kanyesigye, C., Kiwanuka, S., **Kaggwa, C.R.** and Okurut, T.O. 2003. Water quality improvement in the Inner Murchison Bay, Lake Victoria: A case study of algal removal at Gaba II water treatment plant. *The African Journal of Tropical Hydrobiology and Fisheries.* 2: Special Issues. pp 1-22.

Kiwanuka, S., **Kaggwa, R.** Sekayizzi, A. and Okurut T.O. 2002. Municipal wastewater effluent quality improvement and experiences: use of a tropical integrated reconstructed natural wetland in Uganda, Masaka. In: Proceedings of the 8th international conference on wetland systems for water pollution control. IWA, *Comprint International Limited, Dar es Salaam.* 1. Sec. 1-VII pp. 333-337.

Welcomme, R.R., Brummett, R.E., Denny, P., Hasan, M.R., **Kaggwa, R.C**, Kipkemboi, J., Mattson, N.S., Sugunan, V.V. and Vass, K.K. 2006. Water management and wise use of wetlands: enhancing productivity. In: J.T.A. Verhoeven, B. Beltman, R. Bobbink & D.F. Whigham (eds.) *Ecological Studies, Volume 190. Wetlands and natural resource management.* Springer-Verlag, Berlin-Heidelberg, pp. 155-180.

UNESCO-IHE STEP

Name	Rose Christine Kaggwa	**UNESCO-IHE**
Department	Environmental Resources	Institute for Water Education
Period	2001-2006	
Supervisors	Prof. Patrick Denny, Dr. Anne van Dam, Prof. Frank Kansiime	
Daily Advisor	Dr. Anne van Dam	

STEP* ITEM	Year	Study load hours
UNESCO-IHE activities		
PhD seminars (presentation)	2003	40
(participation)	2006	30
Lunch time seminars (6)	2001-2006	180
Other presentations (2)	2005, 2006	60
Colloquium and workshops		
Colloquium (5 Project scientific workshops)	2001-2006	200
Project management workshop (2)	2005	60
International exposure (Conferences & workshops)		
8th international conference on wetland systems for water pollution control, IWA, Arusha, Tanzania, 16th-19th September	2002	30
5th International Conference on Nutrient Cycling and retention in natural and constructed wetlands, Barova Lada, Czech Republic, 24th- 27th September.	2003	30
2nd Symposium for Ecological Sanitation incorporating the 1st IWA specialist group conference on sustainable sanitation, Lubeck, Germany, 7th-11 April.	2003	30
7th INTECOL conference, Utrecht, The Netherlands 25th -30th July	2004	40
Presentations at international forum		
Oral presentation at the 5th International Conference on Nutrient Cycling & retention, Barova Lada, Czech Republic	2003	40
Oral presentation at the 2nd Symposium for ECOSAN Lubeck, Germany	2003	20
Oral presentation 7th INTECOL, Utrecht, The Netherlands	2003	40
In depth studies (additional training)		
Environmental modelling	2006	100
Professional skills (contribution to UNESCO-IHE education programme)		
UNESCO-IHE MSc supervision	2002, 2003	400
UNESCO-IHE education programmes	2003, 2005	30
TOTAL		1330

* STEP (Supervision, Training and Education Plan)

T - #0088 - 071024 - C16 - 254/178/12 - PB - 9780415416979 - Gloss Lamination